U0084892

One Pan for Exotic Meals

One Pan for Exotic Meals

One Pan
for Exotic Meals

*1*個鍋做異國料理

全世界美食一鍋煮透透（中英對照）
World Cuisine in One Pan

洪白陽（CC老師）著

一鍋到底，做菜真簡單！
One Pan for Exotic Meals

在我的廚房和儲藏室裡，堆滿了各式鍋子，有強調功能的、外觀取勝的、每天必用的，以及我的寶貝私藏鍋，我在這些鍋中注入滿滿的情感，視它們如同我的朋友般。因此，也難怪烹飪班的學生和朋友們常揶揄我說：「別人是窮得只剩下錢，我看你呀，是窮得只剩下那些鍋子了。」也就如他們所說，我是個極度的「愛鍋份子」。

我對我的每一個鍋的功能都瞭若指掌，多年來使用得到一個經驗：選對鍋做對菜，絕對能吃得好又省時。而從我的教學經驗中，建議大家可以在家準備好以下幾個常用的鍋子：炒鍋、平底鍋、湯鍋和燉鍋，幾乎什麼菜都能烹調，絕對難不倒你。

此外，親自下廚的人都知道，如果做一頓飯要用到好幾個鍋子，不僅耗費時間，清理更加費力。因此，學生們希望我教大家「只要用一個鍋子」就能完成的料理，所以有了這本書的誕生。我在這本書中選用了炒鍋、平底鍋、湯鍋和燉鍋，告訴大家如何利用這些鍋子完成異國風味的料理，除了海鮮、肉類、蔬菜和湯品，更包括了甜點，讓家裡沒有烤箱的人，也能用鍋子完成美味的糕點。當然並不是要每個人家中都要備妥這幾種鍋子，只要選擇其中幾種，也能料理一桌好菜。

最後，我要感謝一直愛護我的學生、朋友和未曾謀面、卻以信件支持我的讀者們，讓我的烹飪教學生活即使再忙碌，也感到充實而快樂，樂此不疲。期許自己能夠再研發更多美味的中式、異國料理分享給大家，也希望大家都能和我一樣，以料理豐富自己的每一天！

★本次食譜的拍攝，特別感謝助手吳杰修的大力協助！

洪白陽

My kitchen and storage room are filled with different types of pans and pots: some for their function, some for their looks, some I use every day, and some just to round out my collection. I lavish all my love on these pans and pots and treat them as my good friends. No wonder my cooking class students and friends always tease me: "Some people count their wealth in money, but you measure it in pots and pans." And they are right: I am an extreme pan lover.

I know all the functions of my pans. After all these years of using them, my experience is that picking the right pan results in good dishes and saves time. From my teaching experience, I suggest that everyone prepares the following pans and pots: a stir-fry wok, a frying pan, a soup pot, and a stew pot. This collection can be used to cook almost any dish without effort.

People who cook know that preparing a meal can require several pans and pots. This not only wastes time, but the cleanup afterward is a pain. Therefore, my students are always wishing I can teach dishes that can be prepared with "only one pan". That wish was the genesis of this book. Here I show you how to use stir-fry wok, a frying pan, a soup or stock pot, and a stew pot to prepare the most exotic dishes. In addition to seafood, meat, vegetable, and soup dishes, one pan can make delicious desserts even without an oven. Of course, you do not have to prepare all these pans and pots, one or two of them can make a whole table of delicious dishes.

Here I would like to thank my beloved students, friends and readers whom I have never met, yet still support me through their numerous letters. They make my busy teaching life fulfilled and happy. I look forward to cultivating my knowledge of delicious Chinese and foreign dishes to share with everyone. I hope everyone can be just like me, enriching daily life with cooking.

★ Special thanks to the assistance of Wu Jie-shou's help in shooting this cookbook.

Cecilia

目錄 Contents

PART 1 肉類
Meat Section

PART 2 海鮮類
Seafood Section

PART 3 蔬菜和湯類
Vegetable and Soup Section

PART 4 點心和麵包類
Desserts and Breads Section

CC 老師的美食深受台灣、香港等地饕客的喜愛，
只要知道以下幾種食材的香港名稱，
香港讀者也能下廚親嘗！

台灣名稱	香港名稱
西洋芹	西芹
鮭魚	三文魚
貽貝	青口
咕咾肉	咕嚕肉
玉米粉	粟粉
奶油	牛油
鮮奶油	鮮忌廉
乳酪奶油	奶油乳酪
沙拉醬	沙律醬
淡色醬油	生抽

鮮美高湯DIY Delicious Soup Broth DIY

市售的高湯塊、雞湯粉雖然方便，但卻無法得知使用的材料，飲食安全堪慮，建議大家自己燉煮高湯，一次煮大量，分裝後放入冷凍庫保存，使用前再取出解凍。藉由長時間的燉煮，食材的精華全部濃縮在這一鍋高湯之中，搭配任何食材自然無比美味。以下介紹 6 種高湯：基本高湯、雞高湯、牛高湯、日式高湯、泰式高湯、綜合高湯等的做法，基本高湯算是萬用高湯（素食不可），而綜合高湯適合搭配任何海鮮料理。

Soup broth squares or powders sold in markets or shops are convenient, yet there is no way to know what they are made of. To prepare your own food and drink in safety, making your own soup broth at home is highly recommended. Make a big pot of soup, pack into small bags, and store in freezer, then defrost just before use. After boiling for a long time, the essence of the ingredients are concentrated into this big pot of broth. It is prepared with a variety of natural ingredients and is of course delicious and fresh. Six different kinds of soup broth are introduced below: basic soup broth, chicken soup broth, beef soup broth, Japanese soup broth, Thai style soup broth, and combination soup broth. The basic soup broth is the most useful among these (except vegetarian), and the soup broth combination is perfect with seafood dishes.

基本高湯 Basic Soup Broth

材料
雞骨 600 克（1 斤）、豬骨 600 克（1 斤）、西洋芹 2 支、洋蔥 1 個、胡蘿蔔 1 根、月桂葉 1 片、百里香 5 支或乾燥的 1 小匙、水 2,000c.c.

做法
1. 雞骨、豬骨先煎至金黃色（也可用上下火 220 ～ 230℃烤至金黃色）；西洋芹切段；洋蔥、胡蘿蔔都切塊；百里香稍微拍打過。
2. 將所有材料放入燉鍋中，倒入水，熬煮 5 ～ 6 小時，然後過濾即可。

Ingredients
600g chicken bones (1kg), 600g pork bones (1kg), 2 western celery stalks, 1 onion, 1 carrot, 1 bay leaf, 5 stalks thyme or 1t dried thyme, 2000c.c. water

Methods
1. Fry chicken bones and pork bones in pan until golden. (Or bake with upper and lower element at 220~230℃ until golden.) Cut western celery into sections. Cut onion and carrot into pieces. Pat thyme lightly with flat of the knife to enable the flavor to be released more easily.
2. Place all the ingredients in a stew pot along with water, boil for 5~6 hours. Remove and discard the ingredients.

這樣做更省時
Time-saving method

如果使用壓力鍋的話，在做法 **2.** 中只要加入 **1,200c.c.** 水煮，待壓力鍋上升 **2** 條紅線，改小火熬煮 **1** 小時即可。

If a Duromatic is used, in method **2.** only 1200c.c. of water is needed. Cook until the 2 red lines appear, then reduce heat to low and continuing cooking for 1 hour longer.

雞高湯 Chicken Soup Broth

材料

雞骨 600 克（1 斤）、蝦殼 200 克、螃蟹殼 2 隻份量、西洋芹 2 支、小的洋蔥 1 個、胡蘿蔔 1/2 根、百里香 5 支或乾燥的 1 小匙、月桂葉 1 片、白酒 60c.c.、水 7 杯

做法

1. 雞骨先煎至金黃色（也可用上下火 220 ～ 230℃烤至金黃色）；蝦殼以白酒略微炒過；西洋芹切段；洋蔥、胡蘿蔔都切塊；百里香稍微拍打過。
2. 將雞骨、蝦殼、西洋芹、胡蘿蔔和洋蔥、百里香、月桂葉、1 杯水和螃蟹倒入燉鍋，煮至螃蟹熟，撈出螃蟹取出蟹肉放於一旁，螃蟹肉可做其他料理，例如 p.121 的法式鮮蝦蟹肉南瓜湯。
3. 再將螃蟹殼放回燉鍋，倒入剩下的水，熬煮 4 ～ 5 小時，然後過濾即可。

Ingredients

600g (1kg) chicken bones, 200g shrimp shell, 2 portions of crab shell, 2 western celery stalks, 1 small sized onion, 1/2 carrot, 5 stalks thyme or 1t dried thyme, 1 bay leaf, 60c.c. white wine, 7C water

Methods

1. Fry chicken bones until golden. (Or bake in oven with the upper and lower element at 220~230℃ until golden.). Stir-fry shrimp shell with white wine first to eliminate its fishy odor. Cut western celery into sections. Cut onion and carrot into pieces. Pat thyme lightly with flat of the knife to enable the flavor to be released more easily.
2. Combine chicken bones, shrimp shell, western celery, carrot, onion, thyme, bay leaf, 1 cup of water and crabs in stew pot. Cook until the crabs are done, then remove the crab from the pot. Retain the crab meat for other use, such as the French Style Crab and Pumpkin Soup on p.121.
3. Return the crab shell to the stew pot, add the remaining water and continuing cooking for 4~5 hours until done. Remove and discard the dregs.

這樣做更省時
Time-saving method

如果使用壓力鍋的話，在做法 3. 中只要加入 1,200c.c. 水煮，待壓力鍋上升 2 條紅線，改小火熬煮 1 小時即可。

If a Duromatic is used, in method 3. only 1200c.c. of water is needed, cook until there are 2 red lines appear, then reduce heat and cook for 1 more hour.

CC 烹調祕訣 ┃ Cooking tips

1. 本書中 1 杯＝ 225c.c.
2. 雞骨煎烤過味道更香，且可去掉油和雜質
1. 1C=225c.c.
2. The chicken bones taste even better after being fried. Frying helps to reduce oil and any impurities.

牛高湯 Beef Soup Broth

材料

牛骨 1,000 克、丁香 12 根（乾燥香料）、西洋芹 2 支、洋蔥 2 個、胡蘿蔔 1/2 根、大蒜 10 粒、黑胡椒粒 1 大匙、月桂葉 1 片、乾燥百里香 1 小匙、水 2,000c.c.。

做法

1. 牛骨先煎至金黃色（也可用上下火 220 ～ 230℃烤至金黃色）；西洋芹切段；洋蔥、胡蘿蔔都切塊。
2. 將所有材料放入燉鍋中，倒入水，熬煮 6 ～ 8 小時，然後過濾即可。

Ingredients

1000g beef bones, 12 cloves (dried spice), 2 western celery stalks, 2 onions, 1/2 carrot, 10 cloves garlic, 1T black peppercorn, 1 bay leaf, 1t dried thyme, 2000c.c. water

Methods

1. Fry beef bones until golden. (Or bake in oven with the upper and lower element at 220~230℃ until golden.). Cut western celery into sections. Cut onion and carrot into pieces.
2. Combine all the ingredients in the stew pot along with water, cook for 6~8 hours until done. Remove and discard the dregs.

這樣做更省時
Time-saving method

如果使用壓力鍋的話，在做法 **2.** 中只要加入 1,200c.c. 水煮，待壓力鍋上升 2 條紅線，改小火熬煮 1 小時即可。

If a Duromatic is used, in method **2.**, only 1200c.c. of water is needed. Cook until the 2 red lines appear, reduce heat to low and continue cooking for 1 more hour.

日式高湯 Japanese Soup Broth

材料

雞骨 600 克（1 斤）、豬骨 600 克（1 斤）、昆布 1 片、洋蔥 1 個、蘋果 1 個、胡蘿蔔 1/2 根、水 2,000c.c.。

做法

1. 雞骨、豬骨洗淨；洋蔥、蘋果和胡蘿蔔都切塊。
2. 將所有材料放入燉鍋中，倒入水，熬煮 4 ～ 5 小時，然後過濾即可。

Ingredients

600g (1kg) chicken bones, 600g (1kg) pork bones, 1 kombu kelp, 1 onion, 1 apple, 1/2 carrot, 2000c.c. water

Methods

1. Rinse chicken bones and pork bones well. Cut onion, apple and carrot into pieces.
2. Combine all ingredients in stew pot along with water, cook for 4~5 hours, then discard the bones.

這樣做更省時
Time-saving method

如果使用壓力鍋的話，在做法 **2.** 中只要加入 1,200c.c. 水煮，待壓力鍋上升 2 條紅線，改小火熬煮 1 小時即可。

If a Duromatic is used, in method **2.** only 1200c.c. of water is needed, cook until the 2 red lines appear, reduce heat to low and cook for 1 hour.

─── **CC 烹調祕訣** | Cooking tips ───

這道高湯多用在煮湯麵、粥，或者火鍋鍋底。

This soup broth is mostly used in noodle soup, porridge or as the broth in hot pot dishes.

泰式高湯 Thai Style Soup Broth

材料

雞骨 600 克（1 斤）、豬骨 600 克（1 斤）、南薑 5 片、香菜頭 5 個、香茅 3 支、紅蔥頭 1 個、水 2,000c.c.。

做法

1. 雞骨、豬骨先煎至金黃色（也可用上下火 220～230℃烤至金黃色）；香茅切段。
2. 將所有材料放入燉鍋中，倒入水，熬煮 6 小時，然後過濾即可。

Ingredients

600g (1kg) chicken bones, 600g (1kg) pork bones, 5 pieces galangal, 5 coriander roots. 3 lemongrass stems, 1 shallot, 2000c.c. water

Methods

1. Fry chicken bones and pork bones until golden. (Or bake in oven with the upper and lower element at 220~230℃ until golden.). Cut lemon grass stems into sections.
2. Combine all ingredients in stew pot along with water, cook for 6 hours and discard the dregs.

這樣做更省時 Time-saving method

如果使用壓力鍋的話，在做法 **2.** 中只要加入 1,200c.c. 水煮，待壓力鍋上升 2 條紅線，改小火熬煮 1 小時即可。

If a Duromatic is used, in method **2.** only 1200c.c. of water is needed, cook until there are 2 red lines appearing, reduce heat to low and cook for 1 hour.

CC 烹調祕訣 | Cooking tips

這道高湯多用在泰式火鍋的鍋底、湯麵、粥。This soup broth is often used in base of Thai style hot pot, noodle soup, or porridge.

綜合高湯 Soup Broth Combo

材料

雞骨 600 克（1 斤）、豬骨 600 克（1 斤）、蝦殼 300 克、西洋芹 2 支、洋蔥 1 個、小的胡蘿蔔 1 根、月桂葉 1 片、百里香 4 支或乾燥的 1 小匙、水 2,000c.c.。

做法

1. 雞骨、豬骨先煎至金黃色（也可用上下火 220～230℃烤至金黃色）；西洋芹切段；洋蔥、胡蘿蔔都切塊；百里香稍微拍打過。
2. 將所有材料放入燉鍋中，倒入水，熬煮 5～6 小時，然後過濾即可。

Ingredients

600g (1kg) chicken bones, 600g (1kg) pork bones, 300g shrimp shell, 2 stalks western celery, 1 onion, 1 small carrot, 1 bay leaf, 4 stalks thyme or 1t dried thyme, 2000c.c. water

Methods

1. Fry chicken bones and pork bones until golden. (Or bake in oven the upper and lower element at 220~230℃ until golden.) Cut western celery into sections. Cut onion and carrot into pieces. Rub thyme gently between two palms.
2. Combine all ingredients and water in stew pot and cook for 5~6 hours until done, then discard the dregs.

這樣做更省時 Time-saving method

如果使用壓力鍋的話，在做法 **2.** 中只要加入 1,200c.c. 水煮，待壓力鍋上升 2 條紅線，改小火熬煮 1 小時即可。

If a Duromatic is used, in method **2.** only 1200c.c. of water is needed, cook until the 2 red lines appear, reduce heat to low and cook for 1 hour.

認識特殊食材，做菜事半功倍

When you know your special ingredients, the meal is half done

除了常見的食材之外，因為本書中介紹了許多異國料理，所以會用到一些比較特別的食材，像是乳酪、醬料、香料、東南亞食材等，可以到大型或百貨超市、進口食材專門店購買。

In addition to common ingredients that you see all the time, many foreign cuisines and cooking styles are introduced in this book. They use many special, imported ingredients: cheeses, sauces, spices, and Southeast Asian food products. These ingredients you can find in large department store supermarket or a cooking supply store that features imported goods.

A. 香茅（Lemon grass）

在亞洲、泰國等處可見，是烹調許多泰國料理不可或缺的香草，例如：清蒸檸檬魚、酸辣湯等，或者泡製具提神功效的檸檬香茅茶。

You can find it easily in Asia or Thailand. It is a must in Thai dishes such as Lemon Steamed Fish, Sour and Spicy Soup, or Lemon Grass Tea, which has is said to awaken the spirit.

B. 蝦夷蔥（Chive）

又名細香蔥、香蔥，外觀細長，香氣不若一般蔥類嗆辣，味道溫和。

Long and fine in appearance, its aroma is not as strong as that of ordinary scallions. It has a mild taste.

C. 羅勒（Basil）

中式料理中常見的九層塔也屬於羅勒的一種，但味道較羅勒重。羅勒還有其他品種，如茴香羅勒、檸檬羅勒等，多用來燉炒海鮮、肉類、調製沙拉醬和醃泡橄欖油。

The Asian basil often used in Chinese cuisine is stronger than Western sweet basil. Sweet basil comes in many different varieties, including Licorice basil, lemon basil, and Indian basil. It is mostly used in stewing seafood, meats, preparing salad dressings, or pickling in olive oil.

D. 薄荷（Mint）

可用來減少料理的油膩感。新鮮的薄荷可用來泡茶、拌沙拉、搭配醬汁和糕點食用。

Use it to decrease the oiliness of the dish. Fresh mint can be used in making tea, in salads, in salad dressings or in desserts.

E. 百里香（Thyme）

常見於義大利料理中，除了搭配肉類食用，還可以泡茶、釀酒、製作香料橄欖油和製作點心。

It is commonly used in Italian cusine. In addition to cooking with meat dishes, it can be used in tea, wine, flavored olive oil, and dessert.

F. 鼠尾草（Sage）

葉片呈長橢圓形、灰綠色，如同細絨般質感的鼠尾草，多用來搭配海鮮、肉類、蔬菜湯食用。

The leaf is long and oval shaped, grayish green, and has a velvety texture. It is mostly used in seafood, meat or vegetable soup dishes.

G. 迷迭香（Rosemary）

迷迭香的嫩莖、嫩葉除了可以泡茶、泡橄欖油、釀醋之外，可搭配烤雞、烤鴨、烤羊肉等肉類、製作沾醬佐以麵包食用。

In addition to making tea and soaking in olive oil, the tender stem and leaves can be used in roasted chicken, duck or lamb, or in dipping sauces for bread.

H. 洋香菜（Parsley）

又叫巴西里、歐芹、荷蘭芹，全株皆可使用。照片中的是屬於平葉的洋香菜，氣味較溫和，多用來烹調料理。

Also known as Italian parsley, or garden parsley. All stalks can be used in cooking. The parsley shown in the photo is a dark, flat-leaved variety, which has a milder aroma and slender stems with a bright and slightly bitter flavor.

I. 奧勒岡（Oregano）

又叫披薩草、牛至，為義大利料理增添特殊的風味，與披薩、乳酪、蕃茄和橄欖油是最佳的搭檔。通常用來燉肉、燉蔬菜、醃漬、泡茶，用途很廣。

It is used in Italian cuisine to enhance its signature aroma. It is the best friend of pizza, cheese, and tomatoes. It has a wide variety of usages, such as stewing meats and vegetables, flavoring marinades, and making tea.

J. 月桂葉（Bay leaf）

新鮮的月桂葉較乾燥的氣味溫和，多用在燉菜、滷肉，地中海料理很常見。

Fresh bay leaf has a milder taste than dried one. It is often used in stewing vegetables, in meat dishes, and in Mediterranean cuisine.

K. 乾燥百里香（Dried Thyme）

乾燥後的百里香氣味較強烈，使用在烹調上時，用量需調整為新鮮百里香的一半。

Dried thyme has a stronger taste than fresh, reduce the portion by half if used in cooking.

L. 檸檬葉（Lemon Leaf）

又叫卡菲萊姆，是烹調泰式料理不可缺的香料，尤其是加在咖哩中。此外，還用來燉肉、煮湯、醃肉和醃海鮮。

It is also known as Kaffir lime. It is a very important herb in Thai dishes, especially curries. It is also used in stewing pork and in soups as well as in marinating meat and seafood.

蔬果、瓜、蕈類 Vegetables, Squash, Mushrooms

A. 綠捲鬚萵苣生菜
（Oakleaf Lettuce）
又叫法國捲心葉。葉片呈羽毛狀，口感略帶苦味，多用於生菜沙拉或盤飾。
Also called leaf lettuce, the leaves have a feather-like shape, with a slightly nutty and bitter taste. It appears mostly in salads or garnishes.

B. 綠蘆筍（Green Asparagus）
甘脆可口的蘆筍，可去皮後氽燙，涼拌最可口，也可以快炒、煮湯、燒烤。此外，因蘆筍放過久外皮纖維易老化變硬，建議盡快食用。
Crunchy and delicious asparagus can be peeled first, then blanched in boiling water. It is best used in salad. It may also be prepared by stir-frying, in soups, or by barbecuing. Asparagus's skin becomes tougher and coarser if it sits too long. Use it as soon as possible.

C. 芝麻葉（Sesame Leaf）
具有芝麻香的葉子，本身帶點苦味，先用手撕開再烹調，香氣才會散出。可搭配烤肉、湯類食用，但記得絕不可煮太久，以免太苦。
Sesame Leaf: The leaves have a sesame seed smell, yet are slightly bitter in taste. Tear the leaf apart before cooking to release its aroma. It can be paired with barbecued meat and cooked in soups. Remember not to cook too long, or it will become bitter.

D. 紫萵苣生菜（Radicchio Lettuce）
外觀呈球狀，口感清脆多汁，多用於生菜沙拉，搭配沙拉醬就很美味。
Shape like a ball, it has a crunchy and juicy texture. Used in salad, it is delicious just with salad dressing.

E. 東昇南瓜（Dongsheng Pumpkin）

又叫栗子南瓜，外皮呈亮橘色，肉質結實，嘗起來像栗子，口感較 Q，可用來製作甜點，例如本書 p.138 的義大利南瓜乳酪蛋糕。

Also known as chestnut pumpkin. It is light orange on the outside and its texture is firm, yet it tastes like chestnut with a chewy texture. It can be used in making desserts such as the Italian Pumpkin Cheesecake on p.138.

F. 牛蕃茄（Beef Steak Tomato）

又叫陽光蕃茄。久煮不易爛，口感扎實。用來煮湯，湯汁的蕃茄味更濃郁。此外，因含水量較少，可用來夾三明治或漢堡等麵包食用。

Also called beef tomato in the UK. It is firm and tight and not easy to cook through. Using in soup gives it a strong tomato flavor. Because it contains less liquid, it can be sliced and stuffed in sandwiches or hamburgers.

G. 南薑（Galangal）

淺黃棕色，具有淡淡的香氣，是東南亞料理常用到的辛香料。可去除腥味、烹調咖哩、燜煮食物，更能凸顯食材本身的味道。

It has a light brown color and a light aroma. It is very common in Southeast Asian dishes. It can be used to remove the fishy odor of the ingredients. Cooked in curries or simmered dishes, it can bring out the special flavor of the ingredients.

H. 新鮮牛肝菌（Fresh Porcini Mushrooms）

又叫牛薑菇。它是秋季盛產的珍貴菇類，具有香氣，多用在拌炒、燉煮、製作醬汁，或者搭配濃湯、清湯，以增添香氣。

Also known as King Bolete. It is a very precious mushroom only grown in the autumn. Its strong aroma helps in stir-fries, stewed dishes, preparing dressings, or in thick and clear light soups.

I. 青木瓜（Green Papaya）

口感清脆爽口，適合做成涼拌、生菜沙拉、燉煮湯食用，泰國料理中常見的涼拌青木瓜，中式的青木瓜排骨湯都很受歡迎。

It is light and crunchy in texture. Very suitable in making cold salads or in stewing broths. Thai style Cold Green Papaya Salad or Chinese Papaya Rib Soup are very popular.

CC 烹調祕訣 ▍ Cooking tips

乾燥牛肝菌的香氣較新鮮牛肝菌更為濃郁，多以罐裝販售，當次未用完的只要放入冰箱冷藏保存即可。使用前，將牛肝菌放入溫水中泡約 30 分鐘至軟，之後可連同泡牛肝菌的水一起烹調料理。這種歐洲常見的食材，多用在燉飯、炒義大利麵、前菜等。Dried porcini mushroom has a stronger aroma than fresh one. It is mostly sold in cans. Just preserve the leftovers in the can in the refrigerator. Before using, soak the porcini mushrooms in lukewarm water for about 30 minutes until soft, then cook in dish along with the water. This ingredient is the most commonly used ingredient in Europe, mostly used in stewed rice, pasta or appetizers.

罐頭類　Canned Goods Ingredients

A. 紅腰豆（Red Kidney Beans）

營養價值極高的紅腰豆，適合加入肉類、蔬菜中燉煮之外，也可以搭配生菜沙拉、炒飯、炒蔬菜和燉湯。High in nutritional value and tasty in meat and vegetable dishes. May also be used in salads, fried rice, sauted vegetables, and stews and soups.

B. 豬肉豆（Pork and Beans）

以美國土豆（米豆的一種）、豬碎肉、豬油脂、豬肉汁和蕃茄汁等製成的罐頭食品，在歐美等地常食用。可用來燉湯、紅燒、烹調茄汁焗豆等。 High in nutritional value and tasty in meat and vegetable dishes. May also be used in salads, fried rice, sauted vegetables, and stews and soups.

C. 乾蕃茄干（Dried Tomato）

蕃茄乾燥而成，但不同於偏甜的台式乾蕃茄。適合用在燉飯、炒義大利麵、製作醬汁，較油漬乾蕃茄用途多。 Made from tomato dried out in the sun. It is different from Taiwanese tomatoes, which tend to be too sweet. It is suitable in stewed rice, pasta, and sauce. It has more uses than the oily dried tomatoes.

D. 鯷魚（Anchovies）

以橄欖油醃漬新鮮鯷魚，一般多壓成泥製作沙拉醬、沾醬，搭配義大利麵和披薩。市售多以罐裝、盒裝，也有販售加入辣椒的產品。Fresh anchovies pickled in olive oil, mashed and used in salad dressings, dipping sauces, or in sauces for pasta and pizza. You can find canned or packaged ones in the market. Some products even have the chili pepper added.

E. 蕃茄粒罐頭（Whole Tomatoes）

整顆的蕃茄罐頭，通常用在製作蕃茄肉醬、燉肉湯、燉蔬菜。因為已經去除外皮，省時又方便。Cans that contain whole tomatoes. Canned whole tomatoes are used in preparing tomato sauces for meat, stewed meat soups, or braised vegetables. Because the skin its already been removed, they are time saving and convenient to cook.

東南亞食材類　Southeast Asian Products

A. 棕櫚糖（Palm Sugar）

又叫椰子糖，是以棕櫚樹的花汁熬煮而成，東南亞特有的糖類。具有獨特的香氣，甜度比砂糖低，烹調泰式料理時使用，也可加入飲品調味。 Also known as coconut confection. It is made out of liquid gathered from the palm tree. It is a particular sugar only seen in South and Southeast Asia. With its unique aroma, and lower level of sweetness than granulated sugar, it is often used in Thai style cuisine or sometimes in drinks.

B. 酸奶（Sour Cream）

是將牛奶經過發酵處理而成，口感酸濃細滑。除了搭配沙拉、蔬果食用，還可製作優格、加入湯中燉煮，以增添獨特的口味。 It is fermented and processed from milk. With its smooth, sour taste, it is used in salads and vegetables. It can also be made into yogurt. Adding it to stews and soups can increase their unique aroma.

C. 蝦油膏（Shrimp Paste）

泰國獨特的調味料，可用來炒菜、炒飯或當作沾醬，在泰國料理中用途極廣。 It is a special condiment in Thailand. It can be used in sauted vegetables, fried rice, or as dipping sauce. It has a wide range of uses in Thai dishes.

D. 魚露（Fish Sauce）

又叫白醬油，在東南亞料理中等同於我們的醬油，是由小魚、小蝦發酵製成。可代替鹽調味，或用於沾醬、炒青菜。 It is just like soy sauce. It is fermented from small fish and shrimp, and used to substitute for salt. It can be seen in dipping sauce or in sauted vegetable dishes.

E. 椰漿（Coconut Milk）

就是椰奶，不論甜鹹料理都可以加入調味，可用在煮湯、煮咖哩或者製作甜點，為料理帶來特殊的風味。 It can be added to either sweet dishes or salty dishes, or even soups. It brings a special flavor to curries and dessert dishes.

F. 羅望子醬（Tamarind Sauce）

又叫酸子醬，味酸，可加在肉類、海鮮等料理，像名菜泰國酸辣蝦湯，就一定得加入這一味。 It can be added to meats and seafoods. Famous dishes, such as Thai Sour and Spicy Shrimp Soup, cannot be prepared without it.

B. 匈牙利紅椒粉（Hungarian Paparika）

是以紅甜椒去皮乾燥後磨製而成的粉，不具辣味，多用在燉肉、湯品，名菜匈牙利燉牛肉更是非加不可。其香氣和美麗色澤，更可為料理加分。 It is the powder from the dried skinless red bell pepper. It is not spicy and is commonly used in stews and soups. The famous Hungarian Stewed beef cannot be prepared without it. Its beautiful bright color and fragrance adds a lot of points to the dish.

C. 罐裝墨魚醬汁（Squid Ink Jar）

墨魚醬汁多用在墨魚飯、拌炒墨魚義大利麵、西班牙百雅飯中，可為料理增添鮮味和香氣。 Add to risotto, pasta, and Spanish Squid Ink Risotto (arroz negro) to increase the aroma and freshness of the dish.

D. 墨魚醬汁包（Squid Ink Package）

方便使用的小袋包裝，不必擔心用不完的墨魚醬會壞掉，買小袋裝的就能解決問題。Comes in small packs and is convenient to carry around. If you purchase a small pack, you do not have to worry about the leftover squid ink going bad.

E. 第戎芥末醬（Dijon Mustard）

黃色，法式的芥末醬。它是由去掉莢的褐色芥菜籽製成，具辣味，與羊肉、牛肉和豬肉味道很合。除了基本口味之外，另也有加入各種香料，如羅勒、蝦夷蔥、洋香菜等獨特口味的產品。 It is a yellow French style mustard sauce. It comes from the seeds of the brown mustard. It is spicy and perfect with lamb, beef, and pork. In addition to its basic flavor, there are many other unique mustards with different kinds of herbs used as flavorings, including basil, chives, and parsley.

F. 君度橙酒（Orange Liqueur）

又叫香橙酒。由百分之百橙皮提煉，帶有甜橘的自然果香，除了調酒之外，多用來製作糕點、烹調料理。 It is an alcoholic beverage distilled from 100% orange peel. It has that natural fruity aroma. In addition to being used in mixed drinks, it is commonly used in making desserts and in cooking.

醬料和酒類 Sauce and Wine

A. 巴薩米克醋（Balsamic Vinegar）

又叫義大利陳年葡萄醋，具酸甜風味。是以二次濃縮的新鮮葡萄汁在 90℃溫度下熬煮 24 小時，約蒸發掉 1/3 的水分。經過熬煮可以得到具濃郁果香的濃縮葡萄汁，多用在沾醬、調味。 It is sweet and sour, made from fresh concentrated grape juice cooked at 90℃ for 24 hours, with 1/3 of the liquid reduced. It is then fermented through an aging process, leaving the grape juice thick and fruity. It is mostly used in dipping sauces and seasonings.

A. 馬斯卡邦乳酪（Mascarpone Cheese）

以製作提拉米蘇聞名，產於義大利，沒有經過熟成，屬於軟質乳酪，具有淡淡的甜味和奶香，口感細滑。It is an Italian cheese made from cream, coagulated by the addition of citric acid. It is one of the main ingredients in the modern Italian dessert known as Tiramisu, and is sometimes used instead of butter or Parmesan cheese to thicken and enrich risottos. It is a soft cheese and has a light sweet and milky aroma. The texture is very smooth and silky.

B. 奶油乳酪（Cream Cheese）

又叫乳脂乳酪，口感柔滑，屬於軟質乳酪。多用在烘焙，也可烹調料理，或者加入香草製成抹醬，用途很廣。Its soft texture makes it a soft cheese. It is mostly used in baking, or sometimes in cooking. It can also be made into a spread with vanilla added. It has a wide range of uses.

C. 莫札瑞拉乳酪（Mozzarella Cheese）

有經過熟成的半硬質和未經熟成的軟質之分。本書中使用的是半硬質乳酪，而另有一種軟質乳酪，多用在製作前菜，如莫札瑞拉乳酪佐蕃茄青醬。The texture can vary from semi-soft to firm and the flavors from mild and rich to more robust and overpowering. Most of the dishes in this book use semi-hard cheeses. The other soft cheeses are used in preparing appetizers, such as the Mozzarella Cheese with Tomato Pesto.

D. 風提拿乳酪（Fontina Cheese）

屬於半硬質乳酪，是很受歡迎的乳酪鍋材料，經過加熱變得濃稠柔滑，可製作醬汁或當作沾醬。It is a classic Italian cow's milk cheese and is a semi-hard cheese. It is a very popular cheese with hot pot ingredients. Though the heating process, the cheese becomes very thick and smooth. It is used in sauces or dipping sauces.

E. 帕瑪森乳酪（Parmesan Cheese）

依產區使品質略有差異，其中產於 Parmigiano Reggiano 的，有乳酪之王之稱，而 Parmesan Cheese 則最為普通。屬於硬質乳酪，一般多將它刨成片或者磨成粉烹調，用在義大利麵、燉飯等。The quality differs according to the place that produces the cheese. Parmesan from Parmigiano Reggiano has the reputation of the "king of the cheese". Parmesan cheese is a common hard cheese. Ordinarily it is grated into flakes or ground into powder for use in cooking pasta or risotto.

CC 烹調祕訣 ┃ Cooking tips

1. 切乳酪時，乳酪刀需擦乾，不可沾有水分，否則乳酪會發霉。
2. 剩下沒用完的乳酪要用保鮮膜包好，再放入密封保鮮盒中，保鮮盒外再包裹一層濕毛巾，以免乳酪乾掉。
3. 市售乳酪絲可於冷凍保存，現切乳酪則需放在冷藏室。

1. When slicing cheese, the cheese knife has to be absolutely dry, or the cheese will become moldy.
2. Wrap the leftover cheese up with saran wrap, then store in a sealed container. Wrap the container with a wet towel to prevent it from drying.
3. Pre-shredded cheeses sold in packs can be preserved in the freezer. Cheeses which need to sliced before use can be stored in the refrigerator.

實用好鍋與神奇的廚房小幫手
Essential Kitchen Cookwares and Magic Helpers

家中只要備妥一、兩只好鍋，烹調許多料理已經綽綽有餘。這本書中的料理，是用平底鍋、炒鍋、湯鍋和燉鍋這 4 種一般家庭最常用到的鍋子來做的，瞭解每一種鍋子的烹調特性，有助於縮短烹調時間且煮出好吃的料理。

One or two good pans and pots in the kitchen are enough for cooking. In this book, the dishes are prepared with four different kinds of pans that most families use in the kitchen: frying pan, stir-fry wok, soup pot, and stew pot. Understanding each pan's cooking characteristics helps to shorten the cooking time and to prepare a table filled with delicious dishes.

單柄平底鍋

鍋身一體成型，不會藏污納垢、好清洗，且寬口的鍋身設計，除了當炒鍋之外，也能當煎鍋。熱傳導均勻、快速且有相當高的儲熱功能，是省瓦斯節能減碳的好鍋，可無油煎牛排、雞排、羊排。Kuhn Rikon Saute Pan: One piece shaped with an ergonomically designed handle. It does not hide dirt and grease and is easy to clean. With the wide open design it can be used as a saute pan or a frying pan. It has an even and rapid heat distribution. It is energy efficient and carbon reducing. It can fry beef steak, chicken steak, and lamb chops without any oil added.

多層雙耳炒鍋

鍋身一體成型，不會藏污納垢、好清洗，獨特的深鍋設計，炒菜、炒肉不用擔心掉出鍋外。熱傳導均勻、快速且有相當高的儲熱功能，不需使用大火烹煮，食物不會沾鍋，可搭配蒸籠、蒸盤使用。Kuhn Rikon High Dome Wok Multiply Side Grips: One piece molding, with a unique deep wok design, it holds a large amount of food. No food is going to fall out while cooking. With its even and fast heat distribution, there is no need to cook over high heat and the food will not stick to the pan. It can be used with a steamer or a steaming plate.

雙享鍋

可當燉鍋使用。中空斷熱的設計，可有效保有鍋內熱源，傳熱速度非常快，料理的過程也可以全程不洗鍋、不放油直接烹調。Kuhn Rikon Durotherm: It can be used as a stewing pot. With its dual-wall design, it keeps oxygen out and moisture in even when it is off the stove. The rapid heat distribution helps you to create healthy, delicious meals for yourself and your family, using less time and energy. During the process of cooking, no rinsing and no oil is needed, just heat up the food directly.

休閒鍋

分內、外鍋，內鍋使用瑞士頂級不鏽鋼，鍋底能均勻受熱，儲熱性極佳，小火就能煮出美味。而寬口低身的特殊設計，可當炒鍋、平底鍋及湯鍋使用。本書中選用的是 3 公升或 4.5 公升容量的，可當湯鍋。Kuhn Rikon Hotpan: A conventional cookware, made of the finest stainless steel, it is suited for all types of stoves and cooking methods. It stores and distributes heat quickly. Just using low heat can result in a delicious meat dish. Its special low, wide mouthed design allows you to saute and fry, as well as make stews or soups. In this book, the 3 litre or 4.5 litre capacity is used.

CC 特別推薦的
超實用鍋

壓力鍋

適合用來烹調久煮不易熟的食材，只要把全部食材都放入鍋內並計時，待壓力閥下降即可，簡單又省時。以往 5～6 小時才能燉煮出雞精，這個壓力鍋內可快速萃取雞精，讓你 50 分鐘即可上桌享用。此外，8 公升的大容量更可一次烹調 3 道菜，燉肉、煮湯、煮飯一次搞定。Kuhn Rikon Duromatic: Suitable for cooking ingredients that are not cooked easily. Just place all ingredients in the cooker and press the timer until the pressure locking system locks. It is easy and quick. Usually essence of chicken takes 5 to 6 hours to prepare, but with this cooker you can enjoy the dish in only 50 minutes. Moreover, with its 8 Litre capacity, you can prepare three courses, stew, soup, and rice at one time.

易拉轉

免插電的食物調理機，輕輕拉一下，即可將蔥、薑、蒜、辣椒等辛香料迅速攪細，拉的次數愈多越細，從此切洋蔥不流淚，手指也不會殘留蒜味。Kuhn Rikon Swiss Chop Chop: It has a lid with pull-mechanism that rotates blades in the container. Just pull gently and all the strong-smelling ingredients, such as scallions, ginger, garlic and chili pepper, can be chopped finely. The more you pull, the finer it becomes. From now on, there will be no tears from cutting onions and no bad smelling fingers from garlic.

快易夾&精選攪拌湯杓

快易夾和攪拌湯杓是廚房裡最佳的左右手。快易夾可當鍋鏟、打蛋器和夾子。攪拌杓是中空斷熱把手，不會燙手，圓弧的設計，連鍋底死角都能輕易撈出，適用於炒菜及拌湯。Easy-Lock Tong Mediym + Kuhn Rikon Euroline Spoon: Kitchen tongs and stirring ladles are the good right and left hand in the kitchen. The tongs can be used as a spatula, egg beater, and tongs. The stirring ladle has a hole in the center and a soft, heat resistant handle. With its oval design, it can bend to accommodate the edge of the frying pan and reaches every curve in all pots and pans without any effort. It is suitable for frying and for stirring soup.

節能板

可以幫你省下很多瓦斯費的祕密武器！放在瓦斯爐上使鍋子受熱均勻。烹調食物時，節省時間，更可使鍋底不用直接接觸火焰，省去刷洗，同時亦能防止風吹而造成火焰不集中現象，無論小鍋加熱、鋼杯熱牛奶、便當盒熱菜皆宜。Kuhn Rikon Energy saver: It is a secret weapon that reduces electricity bills. For excellent heat conduction and retention, whether it is a small pan, stainless steel cup, or lunch box warming up, it does the job. It protects the outside of pans against discoloration caused by gas flames and saves you much cleaning, and prevents the flame from the flickering caused by breezes.

玉米轉轉樂

玉米造型的外觀設計，新奇又可愛。只要將整支玉米放在防滑底座內，套上外殼，輕鬆轉一轉，隨時隨地都能吃到飽滿的玉米粒，向罐頭玉米粒說再見。Kuhn Rikon Corn Twister: With its corn-shaped appearance, it is new and cute. Just place the corn inside the non-sliding base and secure the outer sheath, twist it gently, and you can have fresh, delicious corn kernels. Say goodbye to canned kernels!

鍋子的清洗
Cleanup Pans and Pots

目前市面上的鍋具依功用、材質而有許多種類產品，以材質來看，最常見的有不沾鍋、不鏽鋼和鑄鐵鍋，其中尤以不沾鍋和不鏽鋼為主。這些我們日常生活中幾乎每天使用的鍋具該怎麼清洗、保養呢？只要掌握以下幾個簡單的原則，鍋子就能長保最佳狀態。

There are many types of cookware in the market, each is different from the other according to the material and function. As for the material, there are non-stick pan, stainless steel and cast iron pan, and non-stick pans and stainless steel pans occupy most of the market. How are we going to clean and maintain these wares that we use every day? Only a few simple principals, pans and pots can be keep at its best condition.

第一次使用新鍋 First Time Using a New Pan

1. 無論不沾鍋或不鏽鋼鍋，先用清潔劑清洗。
   Whether it is a non-sticky pan or a stainless steel pan, use detergent to rinse first.

3. 倒入八分滿的洗米水，或者一般自來水加 2 大匙白醋，然後煮沸。
   Pour rice water in pan up to 80 percent full, or pour in tap water with 2T white vinegar added, then bring to a boil.

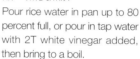

2. 用清水沖洗乾淨。
   Wash pan with water well.

4. 再以清水沖洗乾淨，擦乾鍋子。
   Last, rinse well and then dry with paper towel and store..

不鏽鋼鍋使用後清洗 Stainless Steel Pan Cleanup After Use

1. 趁鍋還熱時清洗，比較容易洗乾淨。
   It is easier to clean the pan when it pan is still hot.

2. 如果忘記熄火，鍋子燒黑紅的話，必須等鍋子冷卻後再慢慢清洗。
   If you forget to turn the stove off and cause a burn-on mess in your pan, wait for the the pan to cool down, then clean slowly.

3. 清洗時，用菜瓜布以畫圓的方式擦洗。
   Clean by scrubbing the bottom of the pan in circles with a scrubbing sponge.

CC 使用不鏽鋼鍋祕訣
Secrets to Prolonging the Life of a Stainless Steel Pan

1. 長期使用的不鏽鋼鍋的鍋壁、鍋底會變黃，建議使用專用清潔劑清洗。
   After a long period of use, the walls and bottom of a stainless steel pan will be stained and covered with a sticky brown mess. Therefore, special detergent for cleaning pans is recommended.
2. 若炒菜時不小心將鍋面燒焦，可先倒入水加熱至微溫再清洗，但若燒焦得太嚴重，可將水煮得再熱一點，並且多等一點時間再清洗。
   If the pan accidentally burns, boil some water inside the pan, then clean. However, if the burnt area is serious, then heat the water up some more, and wait a bit longer to clean the pan.
3. 利用軟性海綿清洗，以免刮壞鍋面。
   Use a soft scrubbing sponge to clean the pan to prevent scratching.

不沾鍋使用後清洗 Non-sticky Pan Cleanup After Use

1. 烹調完畢後不要立刻泡冷水，等稍微降溫再洗，但如果急需立刻使用的話，可先用冷水沖不沾鍋的底部，使鍋子降溫。

   Do not soak the pan in cold water immediately after cooking. Wait until the pan's temperature has decreased. If you need the pan right away, then turn the pan over and rinse the bottom of the pan under cold water to lower the temperature.

2. 翻回鍋面以清水沖洗乾淨。
   Clean pan with water.

3. 用清潔劑清洗，再以清水沖洗乾淨，擦乾鍋子。
   Use detergent to clean first, then rinse well and dry.

CC 使用不沾鍋祕訣
Secrets to Prolonging the Life of a Non-Stick Pan

1. 不要炒帶殼類，例如螃蟹等食材，以免刮傷不沾鍋的塗層。
   Do not use the pan to fry any ingredients with shells, such as crabs, to prevent scratching the coating of the non-stick pan.
2. 避免做蕃茄炒蛋、咕咾肉、糖醋排骨、煮麵、煎鍋貼，因為塗層怕酸鹼性的食材，導致塗層毀壞，但若有標示對應酸鹼質的話則可以使用。
   Avoid making Fried Tomatoes with Eggs, Sweet and Sour Pork, Sweet and Sour Ribs, noodles, and pot stickers, because the coating is vulnerable to acid and alkaline material, which can ruin it. However, if the pan has a label saying that it is alkaline resistant, then it is OK.
3. 避免乾燒，以免鍋子變形或塗層剝落。
   Avoid dry-cooking to prevent the pan from bending out of shape or making the coating peel off.
4. 如果塗層因長時間使用而自然剝落，應該更換新的鍋子。
   If the coating peels off after long use, then it is time to get a new one.

我愛四處旅行，
行經一個又一個陌生的城市，
即使只是漫無目的的行走，
卻總能發現些許耐人尋味且充滿驚奇小事物，
例如，美食！

讓我將旅途中品嘗到的異國風情料理，
也許是燉菜、燉肉、烤肉，又或輕食沙拉，
一道道分享給你。

I love travelling,
one country after another,
even if it is without destination or purpose,
I can always find some intriguing and
surprising little things,
such as delicious food!
Let me share the exotic dishes I have
stumbled across. Whether stewed vegetable
dishes, stewed pork dishes, barbecued pork
dishes, or light salads, all are excellent.

PART I
肉類
MEAT SECTION

威尼斯辣味香煎半雞
Venetian Spicy Chicken

這是一道做法很簡單，聚餐時很受朋友歡迎的料理。利用各式香草和辛香料調味，讓雞肉更顯不同的風味。

This dish is simple to make, it is a very popular dish in parties. Chicken with varieties of herbs and spices is very different from other chicken dishes.

使用這個鍋一鍋到底
平底鍋

材料
雞 1/2 隻、蒜末 1 大匙、紅辣椒末適量、蘑菇 150 克、檸檬 1/2 個、巴薩米克醋 2 大匙、油 1 1/2 大匙

醃料
粗粒黑胡椒粉、鹽（量為 2：1）、迷迭香末 1 1/2 大匙、鼠尾草末 1/2 大匙、百里香末 1 大匙、西洋芹末 2 大匙、洋蔥末 1 1/2 大匙、紅辣椒末適量、蒜末 2 大匙、紅辣椒粉 1/3 大匙、橄欖油 1 1/2 大匙、檸檬汁 1 1/2 大匙

調味料
粗粒黑胡椒粉、鹽適量

做法
1. 將醃料中除了粗粒黑胡椒粉、鹽之外的材料拌勻。
2. 雞清洗後擦乾，先撒上醃料中的粗粒黑胡椒粉、鹽，再均勻地沾上拌勻的醃料，放置 3 小時以上，但建議放入冰箱冷藏一個晚上比較入味。
3. 鍋中倒入 1 大匙油，待油熱後將雞肉皮朝鍋面放入，煎至肉兩面都熟，取出。
4. 鍋中倒入 1/2 大匙油，待油熱後放入蒜末、紅辣椒末爆香，淋入巴沙米可醋，煮至巴薩米克醋僅剩 1 大匙的量，放入蘑菇炒熟，加入調味料拌勻，盛入器皿中，放上雞肉，食用時擠入檸檬汁即可。

Use This Pot
Frying Pan

Ingredients
1/2 chicken, 1T minced garlic, minced red chili pepper as needed, 2T balsamic vinegar, 150g button mushroom, 1/2 lemon, 1 1/2T cooking oil

Marinade
coarsely ground black pepper and salt (proportion 2:1), 1 1/2T minced rosemary, 1/2T minced sage, 1 stalk thyme, 2T minced western celery, 1 1/2T minced onion, minced red chili pepper as needed, 2T minced garlic, 1/3T red chili pepper powder, 1 1/2T olive oil, 1 1/2T lemon juice

Seasonings
salt and coarsely ground black pepper as needed

Methods
1. Combine ingredients from Marinade except coarsely ground black pepper and salt well.
2. Rinse chicken, sprinkle with coarsely ground black pepper and salt first, then coat with method 1. evenly and let sit for over 3 hours. Removing to refrigerator and marinating overnight will enable more of the flavor to be absorbed.
3. Heat 1T cooking oil in pan until smoking, fry chicken with skin facing down until both sides are done and remove.
4. Heat 1/2T of cooking oil in pan until smoking, add minced garlic and red chili pepper, stir-fry until fragrant. Drizzle with balsamic vinegar and cook until 1T of vinegar remains. Add button mushrooms and fry until done, then season with coarsely ground black pepper and salt as needed to taste. Transfer to a serving plate and top with chicken. Squeeze some lemon juice over the chicken. Serve.

CC 烹調祕訣 ┃ Cooking tips

不吃辣的人，可將醃料中的紅辣椒末和紅辣椒粉改成匈牙利紅椒粉，此外，亦可使用雞腿肉製作。

For those who dislike spicy dishes, paprika may be used instead of red chili pepper in the Marinade. Further, chicken legs can be used instead of a half chicken.

法式紅酒燜雞肉

French Style Chicken with Red Wine

一般來說，雞肉大多搭白酒烹調，而這道法式料理則是加入了紅酒、白蘭地，是一道極具成熟風味的雞肉美食。

General speaking, chicken always goes with white wine in cooking. However, red wine and brandy are added to this French dish. It is a dish filled with a mellow flavor.

使用這個鍋一鍋到底
湯鍋

材料
雞腿 3 隻、培根 6 片、洋蔥 1 個、蘑菇 150 克、月桂葉 1 片、西洋芹末少許、紅酒 1 1/2 杯、白蘭地 4 大匙、醬汁 2 杯、奶油 1 1/2 大匙

醃料
粗粒黑胡椒粉、鹽（量為 2：1）

醬汁
雞骨 600 克（1 斤）、豬骨 600 克（1 斤）、蒜頭 10 顆、洋蔥 1 個、西洋芹 2 支、胡蘿蔔 1/2 根、月桂葉 1 片、蕃茄粒罐頭 1 罐、水 1,500c.c.

做法
1. 製作醬汁：雞骨、豬骨先煎或烤至金黃色；洋蔥切塊；西洋芹切段；胡蘿蔔切塊，蕃茄粒搗碎。將雞骨、豬骨倒入湯鍋，加入其他材料，熬煮 4～5 小時，然後過濾即可。
2. 每隻雞腿剁成 2 塊，均勻地撒上醃料，放置 10 分鐘；培根切 1 公分寬；洋蔥切絲。
3. 鍋中倒入奶油，待奶油融化後雞腿皮朝鍋面放入，煎至肉兩面都呈金黃色，取出。
4. 將培根放入煎雞腿的鍋中，煎至焦黃，加入洋蔥，炒至呈透明，再加入蘑菇稍微拌炒。
5. 將雞腿放入做法 4. 中炒均勻，淋入白蘭地煮至酒精揮發，再倒入紅酒煮至酒精揮發，加入月桂葉和醬汁燜煮約 12 分鐘。盛入器皿。

這樣做更省時
Time-saving method

熬煮醬汁時，如果使用壓力鍋，待壓力鍋上升 2 條紅線，改小火燉約 1 小時，最後過濾即可。

If a Duromatic is used to prepare the sauce, cook until the two red lines appear, reduce heat to low and cook for 1 hour, then discard the bones.

Use This Pot
Soup Pot

Ingredients
3 chicken drumsticks, 6 slices bacon, 1 onion, 150g button mushrooms, 1 bay leaf, minced western celery as needed, 1 1/2C red wine, 4T brandy, 2C sauce, 1 1/2T butter

Marinade
coarsely ground black pepper and salt (proportion 2:1)

Sauce
600g chicken leg, 600g pork bone, 10 cloves garlic, 1 onion, 2 western celery stalks, 1/2 carrot, 1 bay leaf, 1 canned whole tomato, 1500c.c. water

Methods
1. To prepare Sauce: First fry or roast chicken and pork bones until golden. Cut onion into chunks. Cut celery into sections. Cut carrot into pieces. Crush whole tomato. Put chicken and pork bone in soup pot, add the rest of the ingredients, simmer and cook for 4～5 hours, then discard the bones.
2. Cut each chicken leg into 2 pieces, sprinkle with Marinade evenly and sit for 10 minutes until flavor is absorbed. Cut bacon into 1cm wide strips. Shred onion.
3. Melt butter in pan, fry chicken legs with skin facing down until golden on both sides, then remove.
4. Fry bacon in the pan just used to fry the chicken legs until golden brown, add onion. Stir-fry onions until transparent, add button mushrooms to mix.
5. Return chicken legs to method 4. and mix well. Drizzle with brandy and cook until the alcohol evaporates, then pour in red wine and cook until the alcohol evaporates once again. Add bay leaf and sauce, simmer for about 12 minutes and transfer to a serving bowl.

CC 烹調祕訣 ▌ Cooking tips

做法 **1.** 的醬汁還可以用在牛排、豬排的沾醬，或者燉肉的高湯醬汁。
The sauce in method (1) can be used as dipping sauce in steak, pork chop, or as soup stock for stewing pork.

凱薩沙拉香煎雞肉

我在常見的凱薩沙拉中，加入了煎烤至香軟的嫩雞肉和獨家的沙拉醬，讓這道輕食變化出新吃法。

Caesar Salad with Fried Chicken

I add fried tender chicken and my special home-made dressing to this ordinary Caesar Salad, and send this dish to new heights!

使用這個鍋一鍋到底
平底鍋

材料

去骨雞腿 3 隻、羅蔓生菜適量、綠捲鬚生菜少許、培根 2 片、法國麵包丁 1/2 杯、帕瑪森乳酪粉適量

醃料

橄欖油 1 1/2 大匙、粗粒黑胡椒粉、鹽（量為 2：1）、檸檬汁 1 大匙、香蒜粉 1 小匙、紅辣椒粉少許

凱薩沙拉醬

鯷魚 5 片、檸檬汁 2 大匙、蛋黃 2 個、鹽適量、黑胡椒粉適量、橄欖油 200c.c.、第戎芥末醬 1 大匙、帕瑪森乳酪粉 3 大匙、鮮奶油 4 小匙

做法

1. 將醃料中除了粗粒黑胡椒粉、鹽之外全部拌勻。
2. 將雞腿先撒上粗粒黑胡椒粉、鹽，再拌入醃料，放置 2 小時。
3. 製作凱薩沙拉醬：將鯷魚放入盆中，倒入檸檬汁，以叉子將鯷魚搗碎，加入蛋黃、鹽、黑胡椒粉，以打蛋器拌勻，接著慢慢倒入橄欖油，邊倒入邊攪拌，最後加入第戎芥末醬、帕瑪森乳酪粉、鮮奶油拌勻即可。
4. 生菜洗淨後放入冰水中冰鎮約 15 分鐘，取出瀝乾水分。培根先切絲，再切成 0.5 公分的小丁，放入鍋中煎至酥脆。
5. 將麵包丁放入鍋中，以小火煎炒至酥脆。
6. 將生菜放入器皿中，擺上雞腿肉，淋上凱薩沙拉醬，撒上培根、麵包丁和帕瑪森乳酪粉即可享用。

Ingredients

3 boneless chicken legs, romaine lettuce leaves as needed, Oakleaf lettuce as needed, 2 slices bacon, 1/2C diced French bread, Parmesan cheese powder as needed

Marinade

1 1/2T olive oil, coarsely ground black pepper and salt (proportion 2:1), 1T lemon juice, 1t garlic powder, red chili pepper as needed

Caesar Dressing

5 slices anchovy, 2T lemon, juice, 2 egg yolks, salt as needed, coarsely ground black pepper as needed, 200c.c. olive oil, 1T dijon mustard, 3T Parmesan cheese powder, 4t whipping cream

Methods

1. Combine the ingredients from marinade except coarsely ground black pepper and salt well.
2. Sprinkle coarsely ground black pepper and salt over chicken leg, then marinate in marinade for 2 hours.
3. To prepare Caesar dressing: place anchovy slices in mixing bowl add lemon juice. Use knife to crush anchovy into small pieces, then add egg yolks, salt and coarsely ground black pepper. Use an eggbeater to beat until evenly mixed, then slowly pour in olive oil while stirring. Last, add Dijon mustard, Parmesan cheese powder and whipping cream and blend well.
4. Rinse lettuce leaves well, soak in ice water for about 15 minutes, then remove and drain. Shred bacon first, then dice into 0.5cm cubes. Fry in pan until crispy.
5. Place bread cubes in pan, fry at low heat until crispy.
6. Place lettuce leaves on a serving plate, top with chicken leg. Drizzle with Caesar dressing, sprinkle bacon, diced bread, and Parmesan cheese powder on top. Serve.

CC 烹調祕訣 | Cooking tips

把鯷魚放入檸檬汁中搗碎，可去除魚腥味，提升香氣。此外，可用豬排肉、牛排或海鮮取代雞腿肉。

Crush anchovy with lemon juice to remove the fishy smell and enhance the aroma. The chicken leg can be replaced with pork chop, beef steak, or seafood.

南法
香煎嫩雞

Southern France Fried Chicken

法式料理也有做法不繁複，人人可嘗
試的，像這道以白酒、白蘭地烹調的
煎嫩雞，只要事先準備好高湯，隨時
隨地都能在家享受浪漫美食。

Some French dishes such as this white wine and
brandy cooked dish are not complicated and
people can easily make them at home. If the soup
broth is prepared beforehand, you can enjoy this
romantic cuisine any time you want.

使用這個鍋一鍋到底
平底鍋

材料

去骨雞腿 3 隻、蘑菇 200 克、紅蔥頭末 2 大匙、蕃茄泥 1/2 杯、基本高湯 1/2 杯（做法參照 p.6）、白蘭地 80c.c.、白酒 120c.c.、玉米粉 1 大匙、洋香菜末少許、奶油 1 大匙

醃料

粗粒黑胡椒粉、鹽（量為 2：1）

調味料

粗粒黑胡椒粉適量、鹽適量

做法

1. 雞腿洗淨後擦乾水分，均勻地撒上醃料，放置 15 分鐘；蘑菇切片。
2. 鍋熱後將雞腿皮朝鍋面放入，煎約 1.5 分鐘後翻面（此時雞皮會呈現金黃色），然後繼續再煎約 2 分鐘，取出。
3. 將奶油倒入剛才煎雞腿的鍋中，待奶油融化後放入紅蔥頭末炒香，續入蘑菇略炒一下，淋入白蘭地煮至酒精揮發，再淋入白酒煮至酒精揮發，倒入蕃茄泥、高湯煮約 6 分鐘，加入調味料拌勻。
4. 加入玉米粉勾芡，然後倒入雞腿煮約 8 分鐘，最後撒些洋香菜末即可。

Use This Pot
Frying Pan

Ingredients

3 boneless chicken legs, 200g mushrooms, 2T minced shallots, 1/2C tomato paste, 1/2C basic soup broth (see methods on p.6 for references), 80c.c. brandy, 120c.c. white grape wine, 1T cornstarch, minced parsley as needed, 1T butter

Marinade

coarsely ground black pepper and salt (proportion 2:1)

Seasonings

coarsely ground black pepper and salt as needed

Methods

1. Rinse chicken well and dry with paper towel, coat evenly with marinade and let rest for 15 minutes. Slice mushrooms.
2. Heat pan first, then place chicken with skin facing down and fry for 1.5 minutes and turn the chicken over (the chicken skin will appear golden), then continue frying for about 2 minutes and remove.
3. Melt butter in the pan and add minced shallots. Stir-fry until fragrant, add mushrooms and stir for a minute or two, drizzle bandy and cook until the alcohol evaporates, drizzle with white wine. Continue cooking until the alcohol evaporates once again, pour in tomato paste and soup broth. Cook for about 6 minutes and add seasonings to taste.
4. Thicken with cornstarch and return chicken. Cook for about 8 minutes until done. Sprinkle with minced parsley and remove from heat. Ready to serve.

CC 烹調祕訣 Cooking tips

1. 蕃茄泥罐頭因本身已有鹹度，在調味時須特別注意。
2. 防止蘑菇片變黑除了加檸檬汁之外，也可在要下鍋前一刻，以快刀切好放入鍋中烹調。
1. Tomato paste already contains salt. Be careful when seasoning to prevent the dish from becoming too salty.
2. Lemon juice can be added to prevent mushrooms from darkening after they are cut open, or you can slice the mushrooms right before cooking.

泰式雞腿佐綠咖哩醬

Thai Chicken Curry

這是一道基本的泰國料理，就像我們的家常菜般，每個泰國家庭都常吃。

This is a basic Thai dish. Just like our homemade dishes, Thai families have it all the time.

使用這個鍋一鍋到底
炒鍋或平底鍋

材料

去骨雞腿 3 隻、綠咖哩 1 1/2 大匙、椰漿
150c.c.、基本高湯或泰式高湯 4 大匙（做
法參照 p.6 和 p.9）、麵粉少許、檸檬葉 5 片、
香茅 1 支、南薑 2 片、九層塔適量

醃料

蒜末 1 1/2 大匙、香菜梗末 1 1/2 大匙、椰漿
3 大匙、咖哩粉 1/2 大匙、薑黃粉（鬱金香
粉）1 小匙、紅蔥頭末 1 大匙、香茅末 1 大
匙、醬油 2 大匙、魚露 1/2 大匙、細砂糖
1/2 大匙、檸檬汁 1 大匙、洋蔥末 1 大匙

做法

1. 雞腿拌入醃料中，放置 2 小時，然後沾上
   麵粉；檸檬葉撕開；香茅切段。
2. 鍋中倒入油，待油熱後將雞腿皮朝鍋面放
   入，煎至肉兩面都熟，取出放在器皿中。
3. 將綠咖哩倒入剛才做法 2. 煎雞肉的鍋中，
   以小火炒至香味溢出，倒入椰漿煮至油浮
   出，注入高湯和檸檬葉、香茅、南薑煮約
   6 分鐘。
4. 將做法 3. 的醬汁淋在雞腿上，以九層塔裝
   飾即可。

Use This Pot

Wok Or Frying Pan

Ingredients

3 boneless chicken legs, 1 1/2T green curry , 150c.c.
coconut milk, 4T basic soup broth or Thai style
soup broth (see methods p.6,9 for references), flour
as needed, 5 stalks lemon leaf, 1 lemongrass stalk,
2 slices galangal, basil as needed

Marinade

1 1/2T minced garlic, 1 1/2T minced cilantro stems,
3T coconut milk, 1/2T curry powder, 1t turmeric
powder, 1T minced shallots, 1T minced lemon
grass, 2T soy sauce, 1/2T fish sauce, 1/2T fine
granulated sugar, 1T lemon juice, 1T minced onion

Methods

1. Marinate chicken legs in marinade for 2 hours
   until flavor is absorbed. Remove and coat
   with flour evenly. Tear lemon leaf apart. Cut
   lemongrass stalk into sections.
2. Heat oil in pan until smoking, fry chicken legs with
   skin facing down until done on both sides and
   place in a serving plate. Retain the oil for later
   use.
3. Pour curry powder to the remaining oil from
   method **2.**, stir over low heat until the flavor is
   released, add coconut milk and cook until the oil
   is floating on the surface. Pour in soup broth, add
   lemon leaf, lemon grass and galangal. Continue
   cooking for 6 more minutes.
4. Drizzle method **3.** over chicken legs and garnish
   with basil leaves. Serve.

CC 烹調祕訣 | Cooking tips

使用檸檬葉要注意，要撕開才會散發出香氣。此外，也可使用豬里肌肉代替雞腿。
**Tear lemon leaf into pieces in order to release its zesty lemon flavor. Pork loin may be used
instead of chicken leg.**

法式香橙烤鴨胸

Duck a l'Orange

法式香橙烤鴨是法國料理中的名菜，我在烹飪課中教過很多次，不過由於學生們對處理整隻鴨非常頭痛，所以用鴨胸取代整隻鴨製作。

Duck a l'Orange is one of the most famous dishes of French cuisine. I have taught it many times in cooking class. However, because the students are not good at handling a whole duck, I use duck breast instead.

使用這個鍋一鍋到底
平底鍋

材料
鴨胸 1 付、油 1/3 大匙

醃料
粗粒黑胡椒粉、鹽（量為 2：1）、香蒜粉
1/2 小匙

醬汁
柳丁（香橙）2 個、檸檬 1 個、細砂糖 2 1/2
大匙、基本高湯 1/4 杯（做法參照 p.6）、
柳丁汁 1/3 杯、鹽適量、玉米粉 1 大匙、君
度橙酒 1 大匙

做法
1. 將醃料拌勻，均勻地撒在鴨胸上，放置 30
   分鐘。
2. 柳丁取出果肉後切小塊，皮切成細絲；1
   個檸檬榨汁，留 1/2 個檸檬皮切成細絲。
3. 鍋中倒入油，待油熱後將鴨胸皮朝鍋面放
   入，煎至肉兩面都熟（也可移入烤箱中，
   以上下火 220℃烤熟），取出切片，排於
   盤中。
4. 製作醬汁：細砂糖倒入鍋中煮至焦糖狀，
   移開火爐，倒入高湯、柳丁皮、檸檬皮 ，
   再移至爐火，倒入檸檬汁煮約 3 分鐘，再
   倒入柳丁汁，加入鹽拌勻，再加入玉米粉
   迅速勾芡，最後淋上君度橙酒，倒入柳丁
   果肉拌勻。
5. 將完成的醬汁倒在鴨胸肉上，鋪上果肉，
   以柳丁皮絲和檸檬皮絲裝飾即可。

Use This Pot
Frying Pan

Ingredients
1 pair of duck breasts,1/3T cooking oil

Marinade
coarsely ground black pepper and salt (proportion
2:1), 1/2t garlic powder

Dipping Sauce
2 oranges, 2 1/2T white granulated sugar, 1/4C
basic soup broth (see methods on p.6 for
reference), 1 lemon, 1/3C orange juice, salt as
needed, 1T cornstarch, 1T cointreau

Methods
1. Combine the ingredients from the Marinade well,
   then sprinkle evenly over the duck breasts and let
   rest for 30 minutes.
2. Cut the oranges open and remove the flesh, then
   cut into small pieces. 1 lemon for juicing, save
   one half and shred peel finely.
3. Heat oil in pan until smoking, fry duck breast
   with the skin facing down first until done on both
   sides. (Or place in oven and bake with upper and
   lower elements at 220℃ until done.) Remove
   duck breasts, slice, then arrange on serving plate.
4. To prepare dipping sauce: Cook white granulated
   sugar in pan until caramelized, remove from heat
   and pour in soup broth, add orange peel and
   lemon peel, then return to heat. Add lemon juice
   and cook for approximately for 3 minutes, then
   pour in orange juice and add salt, then thicken
   with cornstarch. Drizzle with cointreau and add
   orange meat to finish the dish.
5. Pour the sauce over the duck breast and spread
   orange meat on top, then garnish with orange
   peel and lemon peel. Serve.

CC 烹調祕訣 | Cooking tips

刨柳丁皮和檸檬皮時，不可刨到皮白色的部分，否則味道會苦澀。
Avoid the white part of the orange and lemon skin as it tastes quite bitter.

韓式辣味炸雞翅

Korean Spicy Chicken Wings

在我的烹飪教學課中，這一道辣味的炸料理深得各年齡層學生的歡迎，以簡單的做法就能完成不輸餐廳的好菜，難怪每個學生都想學。

In my cooking class this spicy dish is welcomed by students of all ages. With simple methods you can prepare this wonderful dish which is as good as any restaurant dish. No wonder all my students want to learn this dish!

使用這個鍋一鍋到底
平底鍋

材料
雞翅 600 克、麵粉適量、蔥 1 支、熟白芝麻
1 大匙、炸油適量

醃料
醬油 2 大匙、鹽 1/2 小匙、香蒜粉 1/2 小匙

拌料
韓國辣椒醬 3 大匙、醬油 1 1/2 大匙、味醂
1 大匙、香油 1 大匙、薑泥 1 大匙、蔥末 2
大匙、芝麻醬 1 1/2 大匙

做法
1. 雞翅洗淨後瀝乾水分;蔥切絲,然後放入
   冰水中浸泡約 6 分鐘,取出瀝乾水分;將
   拌料拌勻。
2. 將雞翅放入醃料中拌勻,放置 2 小時,然
   後沾上麵粉。
3. 起油鍋,待油溫至 160℃,放入雞翅炸至
   呈金黃色(熟),撈出瀝乾油分。
4. 將雞翅放入拌料中拌勻,取出放在器皿
   中,撒入熟白芝麻和蔥絲即可。

Use This Pot
Frying Pan

Ingredients
600g chicken wings, flour as needed, 1 scallion, 1T cooked/roasted white sesame seeds, deep-frying oil as needed

Marinade
2T soy sauce, 1/2t salt, 1/2t garlic powder

Drizzling Sauce
3T Korean chili paste, 1 1/2T soy sauce, 1T mirin, 1T sesame oil, 1T mashed ginger, 2T minced scallion, 1 1/2T sesame paste

Methods
1. Rinse chicken wings and drain well. Shred scallion. Then rinse and soak in ice water for 6 minutes.Remove and drain. Combine the ingredients from the drizzling sauce and mix well.
2. Marinate chicken wings in Marinade for 2 hours until the flavor is absorbed, then coat wings evenly with flour.
3. Heat oil in pan until smoking, deep-fry chicken wings in oil until golden and done, remove and drain out the oil.Deep-fry inoil at 160℃ until golden,remove to drain.
4. Add chicken wings to the drizzling sauce and mix well. Place in serving plate and sprinkle with white sesame seeds and shredded scallion. Serve.

CC 烹調祕訣 | Cooking tips

1. 蔥絲泡冰水口感才會脆,且較不會辣。
2. 如果沒有芝麻醬,可用無糖花生醬取代。
1. Soak shredded scallion in ice water to remove its spiciness.
2. If sesame paste is not available, use unsweetened peanut butter instead.

法式白酒蛤蜊雞

French Style White Win Clam Chicken

使用這個鍋一鍋到底
平底鍋

材料
帶骨雞腿 2 隻、麵粉適量、培根 2 片、大蒜 3 粒、洋蔥末 1/2 杯、蛤蜊 200 克、貽貝 8 個、洋香菜少許、白酒 150c.c.、基本高湯 1 杯（做法參照 p.6）、油 1 大匙

醃料
粗粒黑胡椒粉、鹽（量為 2：1）

調味料
粗粒黑胡椒粉適量、鹽適量

做法
1. 每隻雞腿剁成 2 塊，撒上醃料，放置 15 分鐘後沾裹麵粉；培根切絲；大蒜切片。
2. 鍋中倒入油，待油熱後將雞腿皮朝鍋面放入，煎至呈金黃色，取出。
3. 將培根倒入煎雞腿的鍋中炒至焦黃，放入蒜片爆香，續入洋蔥末炒香，立刻放入雞腿，淋入白酒煮至酒精揮發，倒入高湯，燉煮約 8 分鐘後加入蛤蜊、貽貝，煮至殼打開，加入調味料拌勻，盛入器皿中，撒上洋香菜即可。這道料理完成後的湯汁不多。

Use This Pot
Frying Pan

Ingredients
2 chicken legs with bone, flour as needed, 2 slices bacon, 3 cloves garlic, 1/2C minced onion, 200g clams, 8 mussels, 150c.c. white wine, 1C basic soup broth (see methods on p.6 for reference), parsley as needed, 1T cooking oil

Marinade
coarsely ground black pepper and salt (proportion 2:1)

Seasonings
coarsely ground black pepper and salt as needed

Methods
1. Chop each chicken leg into 2 pieces and sprinkle with marinade, let sit for 15 minutes, then coat evenly with flour. Shred bacon slices. Cut garlic cloves into slices.
2. Heat oil in pan until smoking, then fry chicken with skin facing down until golden and remove.
3. Fry bacon with the remaining oil from chicken until golden brown, stir-fry garlic until fragrant, then add chicken and drizzle with white wine. Cook until the wine evaporates, pour in soup broth and stew for about 8 minutes. Add clams and mussels, cook until the clam shells open, season with seasonings to taste. Remove to a serving bowl and sprinkle with parsley. There should not be much liquid left after cooking.

白酒蛤蜊義大利麵大家應該很熟悉，這次我要介紹的是一道做法簡單的法國風味白酒蛤蜊雞，蛤蜊吸收白酒後飽滿多汁，搭配鮮嫩雞肉，絕對讓你一口就愛上。

Most people are familiar with White Wine Clam Spaghetti. I would like to introduce this simple French Style White Wine Clam Chicken to my readers. You will fall in love with its juicy, wine-flavored clams and tender chicken.

CC 烹調祕訣 | Cooking tips

蛤蜊本身已有鹹味，所以調味時要先試味道再做調整。

The clams are salty. Taste before seasoning and adjust as necessary.

使用這個鍋一鍋到底
平底鍋

材料
鴨胸 1 付、羅蔓生菜 200 克、洋蔥 1/2 個、紅甜椒 1 個

醃料
味噌 1 大匙、醬油 1/2 大匙、味醂 1/3 大匙、檸檬汁 1 1/2 大匙、蔥末 1 大匙、薑末 1 大匙

柚子沙拉醬
柚子醬 2 大匙、檸檬汁 3 大匙、味醂 1/2 大匙、味噌 1 大匙、橄欖油 120c.c.、柴魚醬油 1 大匙

做法
1. 鴨肉拌入醃料中，放置 2 小時；洋蔥切絲；紅甜椒切條。
2. 製作柚子沙拉醬：將所有材料以打蛋器拌勻即可。
3. 生菜洗淨後放入冰水中冰鎮 15 分鐘，取出瀝乾水分。洋蔥和紅甜椒也分別放入冰水中冰鎮，取出瀝乾水分。
4. 鍋中不放油，待鍋熱後將鴨肉皮朝鍋面放入，煎至肉兩面都熟，取出切片。
5. 將生菜鋪在器皿中，放上鴨肉、洋蔥和紅甜椒，淋上柚子沙拉醬即可享用，可以用洋蔥絲和綠捲鬚生菜裝飾。

和風
鴨肉沙拉
Japanese Duck Salad

Use This Pot
Frying Pan

Ingredients
1 set duck breast, 200g romaine lettuce, 1/2 onion, 1 red bell pepper

Marinade
1T miso paste, 1/2T soy sauce, 1/3T mirin, 1 1/2T lemon juice, 1T minced scallion, 1T mashed ginger

Pamelo Salad Dressing
2T pamelo jam, 3T lemon juice, 1/2T mirin, 1T miso paste, 120c.c. olive oil, 1T bonito flavored soy sauce

Methods
1. Marinate duck meat in marinade for 2 hours. Shred onion and cut red bell pepper in strips.
2. To prepare Pomelo Salad Dressing: Combine all the ingredients and beat with an egg beater until smooth and even.
3. Rinse fresh lettuces and soak in ice water for 15 minutes. Remove and drain. Soak onion and red bell pepper in ice water separately for some time, then remove and drain.
4. Heat pan until smoking first, fry duck meat with skin fcing down without oil until both sides are done. Remove and cut into slices.
5. Place lettuce leaves in serving plate with duck meat on top along with shredded onion and red bell pepper on the side, then drizzle with pamelo salad dressing. Serve.

柚子風味的沙拉醬清爽解膩，一向深受大家的喜愛。這裡用來搭配皮脆肉嫩的鴨肉做成沙拉，顛覆傳統的鴨肉吃法，美味更加分。

This popular pomelo flavored salad dressing is light and clears up the oily feeling. In this salad recipe this dressing is paired with crispy skin duck meat, which subverts the traditional duck serving methods. The delicious flavor is even better.

CC 烹調祕訣 | Cooking tips

也可以使用雞腿肉取代鴨胸肉。
Chicken legs may be used in place of duck.

普利亞橄欖燉嫩雞

Puglia Stewed Chicken with Olives

我在這道義大利料理中添加了橄欖，使食材風味更相互融合，呈現出豐富的味道與口感。

I add olives to this Italian dish to bring out the flavor of the ingredients and give the dish a rich, abundant flavor.

CC 烹調祕訣 | Cooking tips

黑橄欖搭配奶油雞腿非常對味，大家可以試試。

Black olives are perfect with butter chicken leg, I suggest everybody try it.

使用這個鍋一鍋到底
燉鍋

材料
帶骨雞腿 2 隻、麵粉適量、蒜末 1 大匙、洋蔥末 4 大匙、胡蘿蔔末 2 大匙、西洋芹末 2 大匙、黑橄欖 16 顆、白酒 100c.c.、月桂葉 1 片、基本高湯 1/2 杯（做法參照 p.6）、洋香菜少許、奶油 1/2 大匙

醃料
粗粒黑胡椒粉、鹽（量為 2：1）

調味料
粗粒黑胡椒粉適量、鹽適量

做法
1. 每隻雞腿剁成 4 塊，撒上醃料，放置 15 分鐘後沾裹麵粉。
2. 鍋中倒入奶油，待奶油融化後放入雞腿煎至金黃色，取出。
3. 將蒜末倒入煎雞腿的鍋中爆香，續入洋蔥末炒至呈透明，加入胡蘿蔔末、西洋芹末以小火炒約 5 分鐘，倒入雞腿、黑橄欖，淋入白酒煮至酒精揮發，放入月桂葉和高湯燉煮約 15 分鐘，接著加入調味料拌勻，盛入器皿中，撒上洋香菜即可。

Use This Pot
Stewpot

Ingredients
2 chicken legs with bones, flour as needed, 1T minced garlic, 4T minced onion, 2T minced carrot, 2T minced celery, 16 black olives, 8 green olives, 100c.c. white wine, 1 bay leaf, 1/2C basic soup broth (see methods on p.6 for reference), parsley as needed, 1/2T butter

Marinade
coarsely ground black pepper and salt (proportion 2:1)

Seasonings
coarsely ground black pepper and salt as needed

Methods
1. Chop each chicken leg into 4 pieces, marinate in marinade for 15 minutes, then coat evenly with flour.
2. Melt butter in pan, fry chicken leg until golden, then remove.
3. Stir-fry minced garlic until fragrant, add minced onion until transparent, then add minced carrot, minced celery and saute over low heat for about 5 minutes. Return chicken, add black olives to mix. Drizzle with white wine and cook until the alcohol evaporates. Add bay leaf and soup broth and go on cooking for about 15 minutes longer. Season with seasonings to taste. Remove to a serving bowl and sprinkle with parsley. Serve.

使用這個鍋一鍋到底
平底鍋或炒鍋

材料
雞胸肉 300 克、檸檬葉（卡菲萊姆）5 片、蒜末 1
1/2 大匙、香菜梗末 1 大匙、紅辣椒末適量、九層
塔 1 大把、生菜適量、油 1 1/2 大匙、炸油適量

醃料
白胡椒粉 1/2 小匙、淡色醬油 1 大匙、蒜末 1/3 大
匙、檸檬葉末 1/2 大匙

調味料
魚露 1 大匙、細砂糖 1/3 大匙

做法
1. 雞胸肉去皮去骨後切成小碎丁，拌入醃料，放置
   15 分鐘；檸檬葉切絲。
2. 鍋中倒入油，待油熱後放入蒜末、香菜梗末爆香，
   再放入紅辣椒末，倒入雞胸肉、檸檬葉炒至肉快
   要熟，加入調味料拌勻，最後放入半把九層塔快
   炒幾下。
3. 起油鍋，待油溫至 160℃，放入剩下半把九層塔炸
   一下，撈出瀝乾油分。
4. 生菜鋪在盤中，盛入做法 2.，鋪上炸九層塔即可。

Use This Pot
Frying Pan or Wok

Ingredients
300g chicken breast, 5 lemon leaves (Kaffir Lime), 1 1/2T
minced garlic, 1T minced cilantro stem, red chili pepper as
needed, a bundle of basil, lettuce leaves as needed, 1 1/2
cooking oil, deep-frying oil as needed

Marinade
1/2t white pepper, 1T light soy sauce, 1/3 minced garlic,
1/2T minced lemon grass

Seasonings
1T fish sauce, 1/3T fine granulated sugar

Methods
1. Remove skin and bone from chicken breast and dice
   finely, marinate in Marinade for 15 minutes until flavor is
   absorbed. Shred lemon grass.
2. Heat cooking oil in pan until smoking, add minced garlic
   and minced cilantro, fry until fragrant. Add minced red chili
   pepper, then chicken breast and lemon juice, then fry until
   almost done. Add seasonings to taste and throw in a half
   bundle of basil leaves.
3. Heat oil in pan until the temperature reaches 160℃, add
   the remaining basil and deep-fry for a second, remove
   quickly to drain.
4. Spread fresh vegetables on serving plate, top with method
   2. and spread fried basil across top. Serve.

泰北風味肉末

Northern Thai Flavored Minced Chicken with Basil

這是今年泰國朋友介紹給我的北方口味雞肉料
理，美味下飯到令人連吃了好幾盤。我迫不及待
去向老闆討教這道菜的做法，現在分享給大家。
This northern Thailand style chicken dish was introduced
by a friend from Thailand. It was so delicious that I ordered
It again and again. I could hardly wait to ask the boss
how to prepare this dish, and now I'm introducing it to
everyone.

CC 烹調祕訣 | Cooking tips

160℃ 的油溫是指將一塊蔥丟入油鍋
中，會馬上起大油泡泡，一般多用來
炸雞腿、豬排。

**The 160℃ oil temperature may be
measured using a piece of scallion
tossed in the oil. It will release many
bubbles if the oil is at 160C. Most of the
time, it is used for deep-frying chicken
legs or pork chops.**

羅馬尼亞
香煎菇蕈乳酪雞

Romania Cheese Chicken with Porcini

香濃的乳酪裹上雞肉，搭配蕈菇的濃郁香氣，不論是當作主菜、搭配麵條，都是口感多重的超人氣料理。

Chicken wrapped up in cheese with thick rich aromatic Porcini, whether served as a main dish, or noodles, is always a popular dish.

使用這個鍋一鍋到底
平底鍋或湯鍋

材料
去骨雞腿 3 隻、莫札瑞拉乳酪 100 克、大蒜
6 粒、乾燥牛肝菌 60 克、新鮮香菇 6 朵、
帕瑪森乳酪粉適量、油適量、芝麻葉少許

醃料
粗粒黑胡椒粉、鹽（量為 2：1）、橄欖油 1
1/2 大匙、檸檬汁 1 1/2 大匙、香蒜粉 1 小匙、
蒜末 1 1/2 大匙、匈牙利紅椒粉 1/3 大匙、
奧勒岡 1 小匙、油 1/3 大匙

調味料
粗粒黑胡椒粉適量、鹽適量

做法
1. 雞腿洗淨後擦乾，先撒上醃料中的粗粒黑
   胡椒粉、鹽。其他醃料拌勻後放入雞腿，
   放置 2 小時以上。
2. 大蒜切片；牛肝菌放入溫水中浸泡 8 分鐘，
   取出瀝乾水分後切小片；香菇切條。
3. 將雞腿皮朝鍋面放入，先將皮煎至金黃
   色，再將雞腿翻面，以小火慢煎至八分熟，
   （也可移入烤箱中，以上下火 220℃烤至
   八分熟，約 10～15 分鐘），再把莫札瑞
   拉乳酪鋪在雞皮上，蓋上鍋蓋，以小火烤
   至乳酪融化（也可移入烤箱中烤至融化）。
4. 鍋中倒入油，待油熱後放入蒜片爆香，立
   刻放入牛肝菌炒香，續入香菇炒熟，加入
   調味料拌勻。
5. 將做法 4. 盛入器皿中，放上雞腿，撒些
   許現磨的帕瑪森乳酪粉，以芝麻葉點綴
   即可。

Use This Pot
Frying Pan Or Soup Pot

Ingredients
3 boneless chicken legs, 100g mozzarella cheese, 6 garlic cloves, 60g Porcini mushroom, 6 fresh shiitake mushrooms, parmesan cheese powder as needed, cooking oil as needed, rocket salad green as needed

Marinade
coarsely ground black pepper and salt (proportion 2:1), 1 1/2T olive oil, 1 1/2T lemon juice, 1t garlic powder, 1 1/2T minced garlic, 1/3T paprika, 1t oregano, 1/3T cooking oil

Seasonings
coarsely ground black pepper and salt as needed

Methods
1. Rinse chicken legs and dry well, sprinkle with salt and coarsely ground black pepper first, then add the rest of the ingredients from marinade. Marinade for over 2 hours.
2. Cut garlic in slices. Soak porcini mushrooms in lukewarm water for 8 minutes. Remove and drain, then cut in small slices. Cut shiitake mushrooms in strips.
3. Fry chicken legs in pan with skin facing down until golden, then turn legs over and fry over low heat until 80% done. (Or place in oven and bake with upper and lower element at 220℃ on for 10 ~15 minutes until 80% done.)
4. Heat oil in frying pan until smoking, add garlic slices and stir-fry until fragrant, then add porcini mushrooms rapidly until flavor is released, add shiitake mushrooms until done, then season with seasonings to taste.
5. Transfer method 4. to a serving plate, place chicken legs on top. Sprinkle with parmesan cheese powder and garnish with rocket salad greens on the side. Serve.

CC 烹調祕訣 | Cooking tips
1. 如果買不到芝麻葉的話，可用茴香菜或綠捲鬚生菜。
2. 醃肉時加入橄欖油，可使肉變得更軟嫩且增添香味。
1. If rocket salad green is not available, use fennel leaves, or oakleaf lettuce instead.
2. Add a little olive oil to chicken when marinating to soften the texture of the meat and increase the aroma.

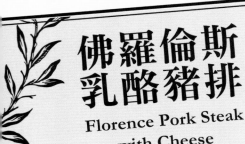

佛羅倫斯
乳酪豬排

Florence Pork Steak
with Cheese

豬排的烹調方式很多，這裡是與柔軟質地且香氣迷人的風提拿乳酪一起煎烤，別有一番滋味。

There are many ways to handle pork chops. In this recipe pork chops are prepared with soft, fragrant Fontina cheese. The results are truly unique.

使用這個鍋一鍋到底
平底鍋

材料
豬里肌肉（切 4 片）600 克、新鮮香菇 8 朵、蘑菇 150 克、風提拿乳酪 100 克、蒜末 1 大匙、洋蔥末 2 大匙、白酒 60c.c.（約 4 大匙）、燉肉的醬汁 4 大匙（參照 p.27）、麵粉適量、蛋液 2 個、帕瑪森乳酪粉 3 大匙、蝦夷蔥末少許、洋香菜葉少許、油 2 大匙

醃料
粗粒黑胡椒粉、鹽（量為 2：1）、香蒜粉 1/4 大匙

調味料
粗粒黑胡椒粉適量、鹽適量

做法
1. 用肉槌將豬里肌肉拍打至鬆軟，撒上醃料，放置 20 分鐘；香菇、蘑菇和風提拿乳酪切片；蛋液和 2 大匙帕瑪森乳酪粉先拌勻。
2. 將豬里肌肉先沾上麵粉，再沾裹拌勻的蛋液。
3. 鍋中倒入 1 1/2 大匙油，待油熱後放入豬里肌肉煎至約七分熟，淋入 2 大匙白酒，煮至酒精揮發，取出。
4. 鍋中倒入 1/2 大匙油，待油熱後放入蒜末、洋蔥末爆香，續入香菇、蘑菇，淋入剩下的白酒，加入調味料拌勻。
5. 將做法 4. 鋪在豬里肌肉上，再排上風提拿乳酪，撒些帕瑪森乳酪粉，蓋上鍋蓋，以小火烤至乳酪融化。
6. 將醬汁淋在平盤上，放上豬里肌肉，撒些蝦夷蔥末和洋香菜葉即可趁熱享用。

Use This Pot
Frying Pan

Ingredients
600g pork tenderloin (cut 4 pieces), 8 fresh shiitake mushrooms, 150g button mushroom, 100g Fontina cheese, 1T minced garlic, 2T minced onion, 60c.c. white wine (approximately 4T), 4T sauce from stewing pork (see methods on p.27 for reference), flour as needed, 2 eggs, 3T parmesan cheese powder, chives as needed, parsely as needed, 2T cooking oil

Marinade
coarsely ground black pepper and salt (proportion 2:1), 1/4T garlic powder

Seasonings
coarsely ground black pepper and salt as needed

Methods
1. Tenderize pork tenderloin with a tenderizer until soft and puffy, sprinkle with marinade and let sit for 20 minutes. Cut button mushroom, shiitake mushrooms, and Fontina cheese in pieces. Combine eggs and 2T of parmesan cheese well together.
2. Coat pork tenderloin evenly with flour, then with egg mixture.
3. Heat 1 1/2T oil in frying pan until smoking. Add pork tenderloin and fry until about 70% done, drizzle with 2T white wine and cook until the alcohol evaporates, and remove from pan.
4. Heat 1/2T oil in pan until smoking. Stir-fry minced garlic and minced onion until fragrant. Then add shiitake mushrooms, button mushrooms, and drizzle in the rest of the white wine. Add seasonings to taste.
5. Return pork and spread method 4. across it evenly, line Fontina cheese on top and sprinkle with parmesan cheese. Cover the pan and cook over low heat until cheese melts.
6. Drizzle sauce over serving plate, place pork tenderloin on and sprinkle with some minced chives and parsely. Serve while still hot.

CC 烹調秘訣 | Cooking tips

1. 蘑菇切片後很容易變黑，可以拌上些許檸檬汁防止變色。
2. 如果買不到蝦夷蔥，可以用洋香菜末取代。
1. The button mushrooms easily darken after being sliced open. Add a little lemon juice to prevent this.
2. If chives are not available, use minced parsley instead.

法式鄉村蔬菜燜烤肉

French Country Veggie with Pork

忙碌的人也想吃好料嗎？那這一鍋就能搞定的超簡單料理一定要推薦給你，不論是宴客或是一般午、晚餐，都能省下不少時間。

Busy people should enjoy good food too; that's why I am recommending this all-in-one frying pan dish to you. Whether entertaining friends or having ordinary lunch or dinner, it saves time and trouble.

使用這個鍋一鍋到底
平底鍋

材料
豬小里肌肉 900 克、培根 150 克、西洋芹 2
支、馬鈴薯 2 個（約 300 克）、胡蘿蔔 300
克、洋蔥 1 個、蘑菇 200 克、奶油 50 克、
大蒜 18 粒、月桂葉 1 片、百里香 1 支、
白酒 120c.c.、基本高湯 1/4 杯（做法參照
p.6）、洋香菜少許、油 1 大匙

醃料
香蒜粉 1/3 大匙、粗粒黑胡椒粉、鹽（量為 2：1）

調味料
粗粒黑胡椒粉適量、鹽適量

做法
1. 豬小里肌肉以棉線綁好固定，防止變形，
   撒入醃料，放置 1 小時。
2. 培根切成小片；西洋芹刨去外層較粗的纖
   維後切小段；馬鈴薯、胡蘿蔔削除外皮後
   都切滾刀塊；洋蔥切塊；奶油切小丁。
3. 鍋中倒入油，待油熱後放入豬小里肌肉煎
   至呈金黃色，放入一半的奶油煎約 1 分鐘，
   取出。
4. 將培根放入剛才煎肉的鍋內，煎至焦黃，
   立即放入 12 粒大蒜、洋蔥、西洋芹略炒
   一下，放入雞腿，淋入白酒煮至酒精揮發
   （約 1 分鐘），倒入剩下的奶油，再加入
   馬鈴薯、胡蘿蔔，蓋上鍋蓋煮至馬鈴薯快
   要熟時，加入蘑菇燜烤至熟，然後瀝出湯
   汁。
5. 將高湯和做法 4. 瀝出的湯汁倒入另一小鍋
   中，放入月桂葉、百里香和剩下的大蒜，
   煮至湯汁濃縮，加入調味料拌勻，再淋在
   做法 4. 的鍋中，撒上洋香菜即可。

Use This Pot
Frying Pan

Ingredients
900g pork tenderloin, 150g bacon, 2 celery strings,
2 potatoes (approximately 300g each), 300g carrot,
1 onion, 200g mushrooms, 50g butter, 18 garlic
cloves, 1 bay leaf, 1 sprigs thyme, 120c.c. white
wine, 1/4C basic soup broth (see methods on p.6
for references), parsley as needed, 1T cooking oil

Marinade
1/3T garlic powder, coarsely ground black pepper
and salt (proportion 2:1)

Seasoning
coarsely ground black pepper and salt as needed

Methods
1. Tie up pork tenderloin with cotton string to
   prevent it from falling apart, sprinkle with
   marinade and let rest for 1 hour.
2. Cut bacon into strips and then into small pieces.
   Peel off outer coarse layer of celery strings and
   cut into small sections. Peel potatoes and carrot
   first, then roll-cut into small pieces. Cut onion into
   chunks. Cut butter into small pieces.
3. Heat oil in pan until smoking, fry pork tenderloin
   until golden, add half amount of the butter and
   continue frying for about 1 minute, then remove
   from pan.
4. With the remaining oil in pan fry bacon pieces
   until brown, add 12 garlic cloves, onion and
   celery. Saute for a minute, add chicken and
   drizzle with white wine and cook for about 1
   minute until the alcohol evaporates, add the
   remaining butter along with potatoes and carrot.
   Cover and cook until the potatoes are almost
   done, add mushroom and simmer until done.
   Drain and retain the liquid.
5. Combine soup broth and liquid from method 4. in
   a small pan, add bay leaf, thyme and remaining
   garlic cloves. Cook until the liquid is thickened,
   season with seasonings to taste, then drizzle with
   method 4. and sprinkle with parsley. Serve.

CC 烹調祕訣 ▍Cooking tips
1. 可選用大里肌肉，但烹調時不宜煎烤得過久。
2. 馬鈴薯削除外皮後必須浸泡冷水，防止變黑，而且可以提升口感。此外，使用新鮮的百里香前，
   先於手掌拍 2 ～ 3 下，可使味道更容易釋出。
1. Pork loin can be used in this dish as well. However, do not fry too long.
2. Soak potato in cold water after the skin is removed to prevent it from darkening. This also helps
   enhance the texture. When fresh thyme is used, pat the thyme between your palms 2 to 3 times
   to release the flavor.

羅馬辣味
豬肉焗麵

Spicy Pork with Penne

焗烤料理一向深受大家歡迎，這裡要介紹一道份量足，又能滿足口腹之慾的焗麵，讀者也可以將豬肉換成其他肉類，美味不減。

Baked cuisine is always welcome at any meal. This baked pasta dish has abundant portions and will satisfy the biggest appetites. Pork can be substituted by other kinds of meat, which will not affect its delicious taste.

使用這個鍋一鍋到底
平底鍋

材料

梅花肉 1,000 克、麵粉適量、蕃茄粒罐頭 1
罐、蒜末 1 大匙、洋蔥末 1 個份量、胡蘿
蔔末 1/4 杯、西洋芹末 1/4 杯、蕃茄糊 1 1/2
大匙、白酒 200c.c.、基本高湯 2 杯（做法
參照 p.6）、月桂葉 1 片、奧勒岡 1/2 小匙、
熟筆管麵適量、乳酪絲適量、帕瑪森乳酪粉
適量、洋香菜末適量、油 2 大匙

醃料

粗粒黑胡椒粉、鹽（量為 2：1）、香蒜粉 1
小匙、紅辣椒粉適量

調味料

粗粒黑胡椒粉適量、鹽適量

炒料

蒜末 1/2 大匙、洋蔥末 3 大匙、紅辣椒末適量

做法

1. 梅花肉切塊，撒上醃料，後放置 20 分鐘，
   然後沾裹麵粉；蕃茄粒搗碎。
2. 鍋中倒入 1 1/2 大匙油，待油熱後放入肉
   塊煎至焦黃，取出。
3. 鍋中直接放入蒜末爆香，續入洋蔥末炒
   香，再放入胡蘿蔔末、西洋芹末炒約 6 分
   鐘，倒入蕃茄糊和煎好的肉塊拌勻，淋入
   白酒煮約 30 秒，然後倒入高湯、蕃茄粒
   和月桂葉、奧勒岡燉煮 40～50 分鐘，至
   肉塊變軟，加入調味料拌勻。
4. 鍋中倒入 1/2 大匙油，待油熱後放入炒料
   爆香，倒入熟筆管麵，加入調味料拌勻。
5. 取適量的做法 3. 和 4 拌勻，倒入烤皿中，
   鋪上乳酪絲，撒些帕瑪森乳酪粉，放入鍋
   中，蓋上鍋蓋，以小火烤至乳酪絲融化（也
   可移入烤箱中，以上下火 220℃烤至呈金
   黃色），撒上洋香菜末即可。

Use This Pot
Frying Pan

Ingredients

1000g shoulder butt pork, flour as needed, 1
canned tomato, 1T minced garlic, 1 onion, 1/4C
minced carrot, 1/4C minced celery, 1 1/2T tomato
paste, 200c.c. white wine, 2C basic soup broth(see
methods on p.6 for references), 1 bay leaf, 1/2t
oregano, cooked penne as needed, shredded
cheese as needed, parmesan cheese powder as
needed, minced parsley as needed, 2T cooking oil

Marinade

coarsely ground black pepper and salt (proportion
2:1), 1t garlic powder, red chili pepper as needed

Seasonings

salt and coarsely ground black pepper as needed

Ingredients for Saute

1/2T minced garlic, 3T minced onion, red chili
pepper as needed

Methods

1. Cut shoulder pork into pieces. Sprinkle with
   Marinade for 20 minutes, then coat evenly with
   flour. Crush whole tomatoes finely.
2. Heat 1 1/2T cooking oil in pan until smoking, stir-
   fry meat until dark brown, then remove.
3. Stir-fry minced garlic until fragrant, add minced
   onion until flavor released, continue adding
   minced carrot and minced celery. Cook for about
   6 minutes, pour in tomato paste and return pork
   to mix. Drizzle with white wine and cook for about
   30 seconds, then pour in soup broth, crushed
   tomatoes, bay leaf, and oregano, then cook
   for 40 to 50 minutes until pork is softened, add
   seasonings to taste.
4. Heat 1/2T oil in pan until smoking. Saute
   Ingredients until flavor is released, add penne and
   seasonings to mix.
5. Mix suitable amounts of method 3. and 4.,
   remove to baking bowl, spread shredded cheese
   evenly across it, then sprinkle with parmesan
   cheese powder. Place in pan with top on, cook
   over low heat until cheese melts. (Or remove to
   oven and bake with upper and lower element at
   220℃ until golden.) Sprinkle with minced parsley
   and serve.

CC 烹調祕訣 ▍Cooking tips

1. 由於平底鍋不具上色效果，可撒上些許匈牙利紅椒粉，視覺上更像焗烤料理。
2. 可以牛肋條或牛腱取代梅花肉製作。

1. Because the frying pan won't brown the dish well, sprinkle with a little paprika to give it the
   appearance of a baked dish.
2. Shoulder pork can be substituted with rib finger beef or beef tendon in this dish.

日式香煎烤肋排

Japanese Fried Ribs

吃膩了美式烤肋排嗎？那這道以日式醬料醃過再烹調的烤肋排，絕對讓你讚不絕口。

Getting tired of American style roasted ribs?
Then this Japanese style marinated fried ribs will
certainly catch your attention!

使用這個鍋一鍋到底
平底鍋

材料

長約 12 ～ 14 公分的肋排 1,000 克、蔥 2 支、紅辣椒 1 條、洋蔥 1/3 個、大蒜 3 粒、昆布 1 片

醃料

醬油 2 1/2 大匙、柴魚醬油 1 大匙、味醂 1 大匙、薑泥 1 大匙、蔥末 2 大匙

烤料

味噌 1 1/2 大匙、醬油 1 大匙、味醂 2 大匙、檸檬汁 1 大匙、蔥末 1 大匙、蒜末 1 大匙、香油 1 大匙、細砂糖 1 小匙

做法

1. 蔥、紅辣椒和洋蔥都切絲；大蒜切片；烤料拌匀。
2. 肋排放入醃料中，放置 1 小時後倒入鍋中，加入洋蔥、昆布和大蒜蒸 1 小時。
3. 取出蒸好的肋排，塗抹烤料後放入鍋內煎烤至焦黃（也可移入烤箱中，以上下火 230℃烤至上色），取出放在盤中。
4. 將蔥、紅辣椒放入冰水中冰鎮約 10 分鐘，取出瀝乾水分，撒在肋排上即可。

Use This Pot
Frying Pan

Ingredients
1000g approximately 12 ～ 14cm long ribs, 2 scallion, 1 red chili pepper, 1/3 onion, 3 garlic, 1 Kombu seaweed

Marinade
2 1/2T soy sauce, 1T bonito flavored soy sauce, 1T mirin, 1T mashed ginger, 2T minced scallion

Barbecue Sauce
1 1/2T miso paste, 1T soy sauce, 2 mirin, 1T lemon juice, 1T minced scallion, 1T minced garlic, 1T sesame oil, 1t fine granulated sugar

Methods
1. Shred scallion, red chili pepper and onion finely. Slice garlic. Barbecue sauce mix well.
2. Marinate ribs in marinade for 1 hour, then place in pan along with onion, Kombu, and garlic added, steam for 1 hour.
3. Remove ribs, spread with barbecue sauce and fry in pan until brown. (Or place in oven and bake with upper or lower element at 230℃ until brown.) Remove to a serving bowl.
4. Soak scallion and red chili pepper in ice water about 10 minutes. Remove and drain, spread evenly over a serving plate, place ribs on top. Serve.

CC 烹調祕訣 Cooking tips

肋排蒸好時湯汁可以留下來，用來煮麵味道極佳。
The liquid released from the ribs when steaming can be retained for later use, such as cooking with noodles. It adds a fantastic flavor!

英國風味燉肉

English Style Stewed Pork

馬鈴薯、洋蔥和肉是西式料理中常見的組合。
具有飽足感的馬鈴薯可直接當作主食，洋蔥則
讓肉與湯汁更加鮮甜。

Potato, onion, and pork are a very popular combination
in western cuisine. The potato filling can be served
directly as the main course. Onions sweeten the pork
and soup broth.

CC 烹調祕訣 | Cooking tips

改用羊肉或牛肉來烹調也很美味喔！

**Lamb or beef, which are also very
delicious, may be used.**

使用這個鍋一鍋到底
燉鍋

材料

豬梅花肉 600 克、洋蔥 600 克、馬鈴薯 600 克、
月桂葉 1 片、酸奶 4 大匙、基本高湯 1 杯（做法參
照 p.6）、洋香菜末少許、香酥的麵包適量、油 1
大匙

醃料

香蒜粉 1 小匙、粗粒黑胡椒粉、鹽（量為 2：1）

做法

1. 梅花肉切成 3X3 公分的塊狀，撒上醃料，放置 30
   分鐘。
2. 鍋中倒入油，待油熱後放入梅花肉，煎至兩面都
   呈金黃色，取出。
3. 洋蔥切塊；馬鈴薯削除外皮後切 1 公分的厚片，
   放入常溫的水中浸泡一下。
4. 將梅花肉、洋蔥和馬鈴薯放入鍋中，倒入高湯，
   擺入月桂葉，燉煮約 60 分鐘至肉變軟，盛入器
   皿中，撒上洋香菜末，搭配 1 大匙酸奶和香酥的
   麵包一起食用。

Use This Pot
Stewpot

Ingredients
600g marble pork, 600g onion, 600g potato, 1 bay leaf, 4T
sour cream, 1C basic soup broth (see methods on p.6 for
reference), parsely as needed, crispy bread as desired, 1T
cooking oil

Marinade
1t garlic powder, coarsely ground black pepper and salt
(proportion 2:1)

Methods
1. Cut pork into 3cm x 3cm big pieces and marinate in
   marinade for 30 minutes.
2. Heat oil in pan until smoking, fry pork until golden brown
   on both sides, then remove.
3. Cut onion into pieces. Peel potates and cut into slices 1 cm
   thick, then soak in lukewarm water for a short while.
4. Place pork, onion and potato in pot, along with soup
   broth, and bay leaf added. Cook for about 60 minutes
   until meat is tender. Remove to a serving bowl and
   sprinkle with parlsey. Serve with 1T sour cream and crispy
   bread on the side.

居酒屋香蔥肉串

Beer House Meat Kababs

下班後和三兩同事到居酒屋吃個肉串,搭配啤酒,令人忘卻一日工作的辛勞。但這些肉串可不便宜,不如在家自己做,經濟又美味。

使用這個鍋一鍋到底
平底鍋

材料
去骨雞腿 3 隻、蔥 3 支、油 1/2 大匙

醃料
味噌 1 1/2 大匙、醬油 2 大匙、香油 1 大匙、味醂 1 大匙、蒜末 2 大匙、蔥末 3 大匙、洋蔥泥 2 大匙、嫩薑泥 1 大匙、胡椒粉 1/2 小匙

做法
1. 蔥切細;醃料事先拌勻。
2. 雞腿肉切塊,拌入醃料中,放置 3 小時,並不時拌一下使肉更入味。
3. 將雞肉塊皮全部朝外,串入竹籤中。
4. 鍋中倒入油,待油熱後將雞腿肉皮朝鍋面放入,先以中火煎,再改小火慢慢煎至肉熟,取出放在器皿中,均勻地鋪上細蔥花即可。

Use This Pot

Frying Pan

Ingredients

3 boneless chicken leg, 3 scallion,1/2T cooking oil

Marinade

1 1/2T miso, 2T soy sauce, 1T sesame oil, 1T mirin, 2T minced garlic, 3T minced scallions, 2T mashed scallion, 1T mashed tender ginger, 1/2t pepper

Methods

1. Cut scallion finely. Combine the ingredients from the marinade well beforehand.
2. Cut chicken into pieces and marinate in marinade for 3 hours until flavor is absorbed. Stir once in a while to enable the flavor to be absorbed more easily.
3. Thread chicken on skewer with skin facing out.
4. Heat oil in pan until smoking, fry chicken skewers with skin face down over medium heat at first, then reduce heat to low and fry until the meat is done. Place in serving plate and sprinkle with finely chopped scallion. Serve.

After work, people always drop by a beer house with their colleagues and order some meat kababs to go with the beer, forgetting the day's work. However, meat kababs are not cheap. Why not make them at home? Here is an economical yet delicious recipe.

CC 烹調祕訣 | Cooking tips

牛肉、豬小里肌肉都很適合用來做烤肉串。

Beef and pork tenderloin are very suitable for making meat kababs.

托斯卡納
野菇蔬菜燉牛肉

Toscana Stewed Beef
with Veggies & Wild Mushrooms

這是一道傳統的義大利料理，適合搭
配義大利麵、法國麵包或白飯食用，
一次燉煮一鍋，輕鬆搞定一餐。

This is a traditional Italian dish, a perfect
accompaniment to pasta, French bread, or white
rice. Just a single pot can solve all the problems of
preparing a whole meal.

使用這個鍋一鍋到底
燉鍋

材料
牛肋條 1,000 克、蕃茄粒罐頭 1 罐、新鮮香菇 6 朵、蘑菇 150 克、培根 5 片、洋蔥末 1 個份量、月桂葉 1 片、蒜末 1 大匙、紅酒 150c.c.、牛高湯 3 杯（做法參照 p. 8）、麵粉 1 小匙、蝦夷蔥或洋香菜末少許、油 1 大匙

醃料
粗粒黑胡椒粉、鹽（量為 2：1）、香蒜粉 1 小匙

調味料
粗粒黑胡椒粉適量、鹽適量

做法
1. 牛肋條切約 4 公分長，撒上醃料，放置 30 分鐘後沾裹麵粉。
2. 蕃茄粒搗碎；香菇和蘑菇切片；培根切 1 公分寬。
3. 鍋中倒入油，待油熱後放入牛肋條煎至焦黃，取出。接著倒入培根炒至焦脆，立即放入蒜末爆香，再倒入洋蔥末炒至呈透明，加入香菇、蘑菇，並倒入煎好的牛肋條，淋入紅酒，煮至酒精揮發。
4. 將蕃茄粒加入做法 3. 中，倒入高湯，放入月桂葉燉煮 30 ～ 40 分鐘，最後加入調味料拌勻，撒上蝦夷蔥末或洋香菜末即可。

Use This Pot
Stewpot

Ingredients
1000g beef rib fingers, 1 can whole tomato, 6 fresh shiitake mushrooms, 150g button mushrooms, 5 slices bacon, 1 onion minced, 1 bay leaf, 1T minced garlic, 150c.c.red wine, 3C beef stock (see methods on p.8 for reference), 1t flour, minced chives or minced parsely as needed, 1T cooking oil

Marinade
coarsely ground black pepper and salt (proportion 2:1), 1t garlic powder

Seasonings
coarsely ground black pepper and salt as needed

Methods
1. Cut beef rib finger into sections about 4cm long, marinate in marinade for 30 minutes, then coat evenly with flour.
2. Crush canned tomatoes. Cut shiitake mushrooms and button mushrooms into slices. Cut bacon into 1cm wide pieces.
3. Heat oil in pan until smoking, fry beef until brown and remove. Next fry bacon until crispy, add minced garlic and fry until fragrant, then add minced onion until transparent. Add shiitake mushrooms, button mushrooms and return beef. Drizzle in red wine. Stew until the alcohol evaporates.
4. Add tomatoes to method 3. to mix, pour in soup broth as well as bay leaf. Stew for 30 to 40 minutes, and season with seasonings to taste. and sprinkle minced chives or minced parsely remove and serve.

CC 烹調祕訣 | Cooking tips
1. 紅酒一定要煮至完全揮發，例如 100c.c. 紅酒揮發濃縮至 50c.c. 的量，不然會苦澀。
2. 麵粉的作用在於增加湯汁的濃稠和使肉質口感軟嫩，但熱量稍高，也可不使用。
1. **The alcohol in the red wine has to be thoroughly evaporated, with the 100c.c.red wine condensed to 50c.c., or the flavor will taste bitter.**
2. **The purpose of adding flour is to thicken the soup and soften the meat. However, it will increase the total calories. Use as desired.**

科西嘉
甜椒燉牛肉

Corsica Stew Beef
with Bell Pepper

這道菜中利用大量的甜椒和牛
蕃茄燉煮牛肉,自然的甘醇和
芳香,不用昂貴的食材,小兵
也能立大功。

This dish uses large quantities of bell pepper and beef
steak tomatoes to stew beef. Without any expensive
ingredients, cheap ingredients with natural sweetness
and aroma can make a big difference.

使用這個鍋一鍋到底
燉鍋

材料
牛肋條 900 克、紅甜椒 400 克、牛蕃茄 500 克、百里香 6 支或乾燥百里香 1/2 大匙、月桂葉 1 片、蒜末 2 大匙、洋蔥末 1 1/2 個份量、麵粉適量、白酒 200c.c.、牛高湯 500c.c.（做法參照 p.8）、紅酒 200c.c.、蝦夷蔥適量、油 3 1/2 大匙

醃料
粗粒黑胡椒粉、鹽（量為 2：1）

調味料
粗粒黑胡椒粉適量、鹽適量

做法
1. 牛肋條切塊，撒上醃料，放置 15 分鐘後沾裹麵粉；紅甜椒和牛蕃茄都去皮後切塊。
2. 鍋中倒入 1 1/2 大匙油，待油熱後放入牛肋條煎至焦黃，淋入白酒煮至酒精揮發，倒入高湯和百里香、月桂葉，煮約 40 分鐘至牛肋條變軟。
3. 另取一個湯鍋，鍋中倒入 2 大匙油，待油熱後放入蒜末爆香，續入洋蔥末炒至呈透明，加入紅甜椒、牛蕃茄炒軟，倒入 1 1/2 大匙麵粉炒勻，熄火。
4. 另取一個小鍋，倒入紅酒煮至酒精揮發，然後倒回做法 3. 鍋中煮 5 分鐘，再整鍋倒入牛肋條鍋中煮 5 分鐘，加入調味料拌勻，放上蝦夷蔥即可。

Use This Pot
Stewpot

Ingredients
900g beef rib finger, 400g red bell pepper, 500g beef steak tomatoes, 6 stalks thyme or 1/2T dried thyme, 1 bay leaf, 2T minced garlic, 1 1/2 onion (minced), flour as needed, 200c.c. white wine, 500c.c. beef stock (see methods on p.8 for reference) ,200c.c. red wine,chives as needed, 3T cooking oil

Marinade
coarsely ground black pepper and salt (proportion 2:1)

Seasonings
coarsely ground black pepper and salt as needed

Methods
1. Cut beef rib finger into sections, marinate in marinade for 15 minutes, then coat with flour evenly. Remove skin from bell pepper and beef steak tomatoes, then cut into chunks.
2. Heat 1 1/2T of oil in pan until smoking, fry beef until dark brown and drizzle with white wine. Cook until the alcohol evaporates, pour in soup broth and add thyme and bay leaf. Continue cooking for about 40 minutes until beef softens.
3. Heat 2T of oil in another soup pot until smoking, saute minced garlic until fragrant, add minced onion and saute until transparent, continue adding red bell pepper and beef steak tomatoes. Wait until tender, fold in 1 1/2T of flour to blend well, then remove from heat.
4. Pour red wine into another small pan and cook until the alcohol evaporates, pour into method 3. and cook for 5 minutes. Pour the whole pot into a beef pot and cook for 5 minutes, season with seasonings to taste and sprinkle with chives. Remove and serve.

CC 烹調祕訣 ▌Cooking tips

紅甜椒和牛蕃茄一定要去皮之後再烹調，以免影響口感。此外，紅酒、白酒也一定要煮至酒精揮發，不然會有苦澀味。
The skin of red bell pepper and beefsteak tomato have to be removed, or it will affect the flavor. The red wine and white wine have to be cooked until the alcohol evaporates, or it will have a bitter taste.

法式酒香燉牛腱

French Style Beef Tendon Stew

在歐式紅酒料理中，這道經典燉菜可以說是我的最愛。濃濃的肉香、軟嫩的口感、味道溫潤而不重。加入具畫龍點睛之效的香草，更顯異國風情。

Among the European red wine dishes, this splendid stew vegetable dish is my favorite. The strong fragrant meat, soft and tender texture, smooth yet not dry flavor, as well as exotic spices and herbs, which makes all the difference.

使用這個鍋一鍋到底
燉鍋

材料
牛腱 1,000 克、培根 5 片、洋蔥 1 個、蘑菇 300 克、
紅酒 180c.c.、醬汁 2 1/2 杯（做法參照 p.27）、
蒜末 1 大匙、月桂葉 1 片、百里香 5 支、蝦夷蔥
末少許、奶油 1 大匙

醃料
粗粒黑胡椒粉、鹽（量為 2：1）、香蒜粉 1 小匙

調味料
鹽、粗粒黑胡椒粉適量

做法
1. 牛腱切約 1 公分厚，均勻地撒上醃料，放置 10
   分鐘；培根切 1 公分寬；洋蔥切末。
2. 鍋中倒入奶油，待奶油融化後放入牛腱，煎至焦
   黃，取出。
3. 將培根放入煎牛腱的鍋中，煎至焦黃，加入洋
   蔥、蒜末炒至洋蔥呈透明，再加入牛腱、蘑菇
   拌勻，淋入紅酒煮至酒精蒸發，倒入醬汁。
4. 百里香放在手掌心拍兩下，連同月桂葉一起加入
   做法 3. 中，燉煮約 50 ～ 60 分鐘，加入調味料，
   撒入蝦夷蔥末即可。

這樣做更省時
Time-saving method

燉煮時，如果使用壓力鍋，待壓力鍋上升 2
條紅線，改小火燉約 23 ～ 25 分鐘，最後
加入調味料即可。

If a Duromatic is used to prepare this dish, cook
until the two red lines appear, reduce heat to low
and cook for 23 ～ 25 minutes, then season with
seasonings to taste.

Use This Pot
Stewpot

Ingredients
1,000g beef tendon, 1 onion, 5 slices
bacon, 300g mushroom, 180c.c. red wine,
2 1/2 cups sauce (see methods on p.27 for
reference), 1T minced garlic, 1 bay leaf, 5
stalks thyme,chives as needed, 1T butter

Marinade
coarsely ground black pepper and salt
(proportion 2:1), 1t garlic powder

Seasonings
salt and coarsely ground black pepper as
needed

Methods
1. Cut beef tendon into slices about 1cm
   thick, sprinkle evenly with ingredients from
   Marinade, and marinate for 10 minutes.
   Cut bacon into 1cm wide pieces , and
   mince onion.
2. Heat butter in pan until it melts, add beef
   tendon and fry until light brown, then
   remove.
3. Put the bacon in the pan that was just
   used to fry the beef tendon and fry until
   light brown. Add onion and minced
   garlic, fry until transparent. Return beef
   tendon and mushrooms to mix. Drizzle
   with red wine and cook until the alcohol
   evaporates, then pour in sauce.
4. Place thyme between palms and rub
   twice, add to method 3. along with bay
   leaf, stew for about 50～60 minutes. Add
   seasonings and sprinkle minced chives to
   taste. Serve.

CC 烹調祕訣 Cooking tips

1. 牛腱燉煮的時間需視牛腱的筋而定，筋愈多價格愈貴，燉煮的時間也愈長，可自行斟酌，而
   澳洲牛腱燉煮的時間則較短。
2. 紅酒的酒精一定要煮至完全揮發，不然醬汁會有苦澀味。

1. **The cooking time for preparing tendon depends on the amount of beef tendon: the greater the
   amount, the longer the cooking time. For Australian beef tendon the cooking time is shorter.**
2. **The alcohol in the red wine has to be completely evaporated, or the sauce will taste bitter.**

印度咖哩燉羊肉

Indian Curry Stewed Mutton

最怕羊腥味的我通常只吃小羔羊（法式羊排），但這道羊肉料理不僅沒有一丁點腥味，加入異國風味的香料烹調，絕對讓人一吃上癮。

I strongly dislike the gamey odor of mutton and usually eat only lamb. Yet this dish does not taste smelly bit gamey. Flavored with exotic spices, this dish will make you addicted.

使用這個鍋一鍋到底
燉鍋

材料

羊肉 1,000 克、洋蔥 1 個、牛蕃茄 2 個、蒜末 2 大匙、薑末 1 大匙、香菜籽 5 克、咖哩粉 2 大匙、基本高湯 2 杯（做法參照 p.6）、無糖優格 150c.c.、香菜適量、鹽適量、油 1 1/2 大匙

醃料

無糖優格 100c.c.、醬油 2 大匙、咖哩粉 1 大匙、薑黃粉 1 小匙、蒜末 1 大匙、薑末 1 大匙、白胡椒粉 1 小匙、鹽 1 小匙

香料

肉桂粉 1/2 小匙、小豆蔻 5 粒、丁香 4 粒、小茴香籽 1/2 小匙、月桂葉 1 片、黑胡椒 1/4 小匙

調味料

鹽適量

做法

1. 將羊肉放入醃料中拌勻，放置 1 小時；洋蔥切末；牛蕃茄去皮。
2. 鍋中倒入 1 大匙油，待油熱後放入羊肉，煎至兩面都呈金黃色，取出。
3. 將 1/2 大匙油倒入煎羊肉的鍋中，待油熱後先放入蒜末、薑末爆香，續入洋蔥末炒至呈透明，再加入香菜籽、咖哩粉，以小火炒至咖哩香味散出，放入羊肉拌勻。
4. 倒入高湯，加入所有香料、牛蕃茄，燉煮至肉變軟，再倒入優格煮約 3 分鐘，加入調味料，盛入器皿內，撒上香菜，可搭配白飯或印度煎餅食用。

Use This Pot
Stewpot

Ingredients

1000g mutton, 1 onion, 2 beef steak tomatoes, 2T minced garlic, 1T minced garlic, 5g cilantro seeds, 2T curry powder, 2C basic soup broth (see methods on p.6 for reference), 150c.c. plain yogurt no sugar added, cilantro and salt as needed, 1 1/2T cooking oil

Marinade

100c.c. plain yogurt (no sugar added), 2T soy sauce, 1T curry powder, 1t turmeric powder, 1T minced garlic, 1T minced ginger, 1t white pepper, 1t salt

Spices

1/2t cinnamon powder, 5 small cardamom, 4 cloves, 1/2t cumin seeds, 1 bay leaf, 1/4t coarsely ground black pepper

Seasonings

salt as needed

Methods

1. Marinate mutton meat in Marinade for 1 hour. Chop onion finely. Peel off skin from beef steak tomatoes.
2. Heat 1T cooking oil in pan until smoking, fry mutton until golden on both sides, then remove.
3. Add 1/2T oil in the pan from Method 2., heat until smoking, stir-fry minced garlic and ginger until fragrant, then add minced onion and fry until transparent. Add cilantro seeds and curry powder, stir over low heat until curry flavor is released, return mutton.
4. Pour in soup broth, along with all the spices and tomatoes. Stew until the meat is softened, then pour in yogurt and cook for 3 minutes longer. Season with seasonings to taste and transfer to a serving bowl, then sprinkle with cilantro on top. Serve with white rice or Indian naan.

這樣做更省時
Time-saving method

燉煮時，如果使用雙享鍋的話，只需加入 1 杯高湯即可。

Only 1C of soup broth is needed if prepared using a Durotherm.

CC 烹調祕訣 | Cooking tips

1. 這道料理除了可用羊肉以外，牛肉、豬梅花肉也很適合。
2. 因每個品牌的咖哩粉配方不同，建議將 3 種以上不同品牌的咖哩粉混合後再使用，咖哩香味更提升。

1. Besides mutton, this dish can be prepared with beef or shoulder butt pork, they are quite suitable.
2. Due to different brand of curry powder contains different ingredients, using the combination of over 3 different brands of curry powder together is suggested to upgrade the dish.

布根地
紅酒醬汁牛排

Burgundy Beef Steak
with Red Wine Sauce

在我教過的多道牛排料理中，這道是很受學生歡迎的一道，尤其用來宴客招待朋友，更受到熱烈的迴響，美味可媲美五星級餐廳。

This is one my students' favorite steak dishes. When they use this to entertain friends or guests, they always receive enthusiastic applause. Its delicious taste would do justice to a five star restaurant.

使用這個鍋一鍋到底
平底鍋

材料
8 盎司菲力牛排 4 塊、培根 4 片、蘆筍 100
克、紫萵苣、綠捲鬚生菜和紅甜椒絲少許、
紅酒 120c.c.、奶油 1 1/4 大匙

醃料
粗粒黑胡椒粉、鹽（量為 2：1）

濃縮汁
牛高湯 1/4 杯（做法參照 p.8）、醬汁 1/4
杯（做法參照 p.27）

調味料
鹽、粗粒黑胡椒粉適量

做法
1. 將菲力牛排圍上培根，以牙籤固定好，均
   勻地撒上醃料，放置 5 分鐘；蘆筍放入滾
   水中汆燙熟後取出。
2. 製作濃縮汁：將牛高湯和醬汁倒入鍋中，
   煮至剩約 1/4 杯即成。
3. 鍋中倒入 100c.c. 紅酒，煮至剩 50c.c.，
   再倒入濃縮汁，煮至剩約一半的量，加入
   1/4 大匙奶油，倒入調味料拌勻。
4. 鍋中倒入 1 大匙奶油，待奶油融化後放
   入牛排，煎至喜愛的熟度（培根要煎至
   焦黃）。
5. 將做法 3. 倒入盤中，放上蘆筍，再擺入牛
   排，以紫萵苣、綠捲鬚生菜和紅甜椒絲點
   綴即可。

Use This Pot
Frying Pan

Ingredients
4 pieces 8oz fillet steak, 4 slices bacon, 100g
asparagus, okaleave lettuce, romaine lettuce leaves
and red sweet pepper as needed,120c.c. red wine,
1 1/4T butter

Marinade
coarsely ground black pepper and salt (proportion
2:1)

Gravy
1/4C beef stock (see methods on p.8 for reference),
1/4C sauce (see methods on p.27 for reference)

Seasonings
salt and coarsely ground black pepper as needed

Methods
1. Wrap steak up with bacon slices, secure with
   toothpicks and sprinkle evenly with ingredients
   from marinade, then marinate for 5 minutes.
   Blanch asparagus in boiling water until done and
   remove.
2. To prepare Gravy: Heat beef stock and sauce in
   pan until approximately 1/4C remains.
3. Heat 100c.c. red wine until only 50c.c. remains,
   add gravy, cook until half remains. Add 1/4T of
   butter, then add seasonings to mix well.
4. Heat 1T of butter until melts, fry steak done as
   desired. (Bacon has to be crisped.)
5. Pour method 3. on serving plate, top with
   asparagus, then place steak and sprinkle
   okaleave lettuce, romaine lettuce leaves and red
   sweet peppe on the side. Serve.

CC 烹調祕訣 ▏Cooking tips

1. 可利用 p.59 法式酒香燉牛腱的湯汁代替濃縮汁。
2. 牛排熟度的簡易測試方法，三分熟類似耳垂的柔軟度，六分熟類似臉頰的柔軟度，八分熟則
   類似鼻頭的柔軟度。

1. Gravy can be substituted with liquid from p.59 French Style Wine Beef Tendon Stew.
2. The easy way to test the steak's doneness: 30% done is similar the softness of an earlobe, 60%
   done is similar to softness of a cheek, 80% done is similar to the softness of the nose.

普羅旺斯
烤羊排佐甜椒醬

**Provence Roasted Lamb Chops
with Bell Peppers**

將香料搭配醃過的羊排，再佐以馬
鈴薯甜椒醬，如此特別的吃法，喜
愛烤羊排的你一定沒嘗過，馬上來
試試吧！

Herb marinated lamb chops paired with potato and
bell pepper sauce. I bet lovers of lamb chops will
have never tasted this special dish. Why not try it?

使用這個鍋一鍋到底
平底鍋

材料
羊排 12 支、第戎芥末醬 4 大匙

醃料
粗粒黑胡椒粉、鹽（量為 2：1）、新鮮百里香末 2 大匙或乾燥百里香 2/3 大匙、迷迭香末 3 大匙、蒜末 2 大匙、洋香菜末 1 1/2 大匙、香草適量

沾裹料
羅勒末 3 大匙、麵包粉 4 大匙、蒜末 2 大匙

沾醬
蒸爛的馬鈴薯 1 個、炒熟的紅甜椒 1 個、蒜末 1 大匙、鮮奶油 3 大匙、牛高湯 3 大匙（做法參照 p.8）、鹽適量

做法
1. 將醃料中除了粗粒黑胡椒粉、鹽之外全部拌勻。
2. 羊排先撒上粗粒黑胡椒粉、鹽，再拌入醃料，放置 2 小時。
3. 將羊排放入平底鍋中煎至呈金黃色（不可太熟），取出塗抹少許第戎芥末醬，沾上沾裹料，放入平底鍋，以小火煎至熟（也可移入烤箱中，以上下火 220℃烤熟）。
4. 製作沾醬：將所有材料放入調理機中打成泥狀即可。
5. 將沾醬淋在器皿中，放上羊排，以香草點綴即可。

Use This Pot
Frying Pan

Ingredients
12 lamb chops, 4T Dijon mustard

Marinade
coarsely ground black pepper and salt (proportion 2:1), 2T fresh minced thyme or 2/3T dry thyme, 3T minced rosemary, 2T minced garlic, 1 1/2T minced parsley, herb as needed

Coating
3T minced basil, 4T bread crumbs, 2T minced garlic

Dipping Sauce
1 steamed and mashed potatoes, 1 fried red bell pepper, 1T minced garlic, 3T whipping cream, 3T beef soup broth (see methods on p.8 for reference), salt as needed

Methods
1. Combine the ingredients from marinade except coarsely ground black pepper and salt, then mix well.
2. Sprinkle coarsely ground black pepper and salt on lamb chop, then stir in marinade. Marinate for 2 hours.
3. Fry lamb chops in pan until golden (do not overcook), remove and spread evenly with a little Dijon mustard, then coat evenly with bread crumb coating and return to frying pan. Fry over low heat until done. (Or place in an oven and bake with upper or lower element at 220℃ until done.)
4. To prepare dipping sauce: Put all the ingredients in blender and blend until mushy.
5. Drizzle the dipping sauce on a serving plate and top with lamb chop. Sprinkle with herb and Serve.

CC 烹調祕訣 ▎Cooking tips

1. 沾醬還可以用來沾食海鮮和法國麵包。
2. 乾燥香料的用量是少於新鮮香料的 1.5 ～ 2 倍。

1. The dipping sauce can be served with seafood and French bread.
2. The amount of dried herbs should be 1.5~2 times less than fresh.

墨西哥風味
茄汁燉牛肉

Mexican Stewed Beef with Tomato Sauce

喜歡吃肉的人絕對不能錯過這道彌漫南美風情的異國料理,尤其在冬天,濃厚的肉香更能溫暖脾胃。

Meat lovers should not miss this South American style dish. Its thick meaty aroma will warm you're your stomach, especially in the winter.

使用這個鍋一鍋到底
燉鍋

材料
牛肋條600克、豬胛心肉200克、蕃茄粒罐頭1罐、培根絲3大匙、蒜末2大匙、洋蔥末1個份量、奧勒岡1/3大匙、羅勒葉末2大匙、紅腰豆1罐、紅酒150c.c.、牛高湯400c.c.(做法參照p.8)、香菜少許、油1大匙

醃料
粗粒黑胡椒粉、鹽(量為2:1)、紅椒粉1大匙、香蒜粉1/4大匙

調味料
粗粒黑胡椒粉適量、鹽適量

做法
1. 牛肋條、豬胛心肉都切塊,撒上醃料,放置30分鐘;蕃茄粒搗碎。
2. 鍋中倒入油,待油熱後放入牛肋條、豬胛心肉,煎至兩面都呈金黃色,取出。
3. 將培根倒入煎肉的鍋中炒至焦黃,放入蒜末爆香,續入洋蔥末炒至透明,再放入牛肋條、豬胛心肉,淋入紅酒煮至酒精揮發,加入蕃茄粒、奧勒岡、羅勒葉末、紅腰豆和高湯,燉煮40～50分鐘至肉變軟,以香菜點綴。

Use This Pot
Stewpot

Ingredients
600g beef rib finger, 200g pork shoulder, 1 canned whole tomato, 3 shredded bacon, 2T minced garlic, 1 minced onion (minced), 150c.c. red wine, 1/3T oregano, 2T minced sweet basil, 1 canned red kidney beans, 400c.c. beef soup broth (see methods on p.8 for reference), cilantro as needed, 1T cooking oil

Marinade
coarsely ground black pepper and salt (proportion 2:1), 1T paprika powder, 1/4T garlic powder

Seasonings
coarsely ground black pepper and salt as needed

Methods
1. Cut beef rib finger and shoulder pork into chunks, marinate in marinade for 30 minutes. Crush whole tomato finely.
2. Heat cooking oil in pan until smoking. Fry beef rib fingers and shoulder pork in pan until both sides are golden, then remove.
3. Fry bacon in pan from method 2. until light brown. Stir-fry minced garlic until fragrant, add minced onion and cook until transparent. Next, return beef rib finger and shoulder pork, drizzle with red wine and simmer until the alcohol evaporates. Add crushed tomato, oregano, and red kidney beans as along with the soup broth. Stew for 40 to 50 minutes until meat is soft and tender. Ready to serve.

印度煎餅
Indian Naan

一般市面上很難買到印度煎餅，建議大家自己嘗試製作！

Indian Naan is rarely seen in stores and markets. I suggest that you try making it at home.

使用這個鍋一鍋到底
平底鍋

Use This Pot
Frying Pan

材料
中筋或全麥麵粉 200 克、鹽 1/2 小匙、水 180c.c.、室溫軟化的奶油適量、油適量

做法
1. 中筋麵粉過篩後倒入盆內，麵粉中間做成一個凹槽，放入鹽，然後慢慢倒入水。
2. 將旁邊的粉慢慢撥進來，把全部材料混合均勻，搓揉成光滑的麵團，然後用保鮮膜包好麵團，放置 30 分鐘。
3. 取出麵團，平均分割成 8 ～ 10 份，每一份擀成圓片狀，塗上奶油後對折再擀平，接著再次塗上奶油後再對折，擀平成類似三角形。
4. 鍋中倒入適量的油，待油熱後放入三角餅皮，煎至餅皮兩面都熟即可。
5. 可搭配 p.61 的印度咖哩燉羊肉一起食用。

Ingredients
200g all purpose or whole wheat flour, 1/2t salt, 180c.c. water, butter (soften at room temperature) as needed

Methods
1. Sift all purpose flour first then fold into pan, form a hole in the center and add salt. Slowly pour in water.
2. Move the flour from the side to center slowly, mix all the ingredients, then knead into a smooth soft dough, then wrap up with saran wrap and sit for 30 minutes.
3. Remove dough, cut evenly into 8 ～ 10 equal pieces. Roll into a round circles, spread butter evenly on surface and fold up, roll into a round again, then spread butter over and fold up, roll into a round once again into a triangular shape.
4. Heat oil in pan until smoking, fry triangular dough until done on both sides.
5. It may be served with Indian Curry Stewed Mutton on p.61.

一年四季，
都是吃美食的好時節。

在朝氣的新春，
享受清爽的利波諾海鮮沙拉；
在無風的夏夜，
來杯啤酒佐越南風味拌墨魚；
在涼爽的初秋，
大啖肥美的地中海風味鮭魚；
在凜冽的冷冬，
品嘗暖脾胃的葡萄牙海鮮鍋。

Each season of the year offers its own culinary delights.
In the vibrant spring, enjoy Lipno Seafood Salad. On a sweltering summer night, have a glass of beer with Vietnamese Squid Cold Dish. On a cool autumn day, enjoy fatty, delicious Mediterranean Style Salmon. In the cold of winter, warm yourself with Portugese Seafood Pot.

PART 2
海鮮類
SEAFOOD SECTION

米蘭香料魚排

Milan Herbal Fish Steak

這是一道充分運用新鮮香草烹調成的西式料理。假日的餐桌上,一家三口悠閒地品嘗西式晚餐,伴隨著笑語,再愜意不過了。

This is a western dish, which is prepared thoroughly with fresh herbs. During the holidays, a family of three can casually enjoy a western meal, chatting and laughing their way through the meal. Nothing could be more delightful.

使用這個鍋一鍋到底
平底鍋

材料

鰈魚、鯛魚或鮭魚 3 片（約 600 克）、四季豆 100 克、麵粉 4 大匙、蛋液 2 個份量、奶油 1/2 大匙、蒜末 1/2 大匙、油 2 1/2 大匙

醃料

白酒 2 大匙、粗粒黑胡椒粉、鹽（量為 2：1）

沾裹料

麵包粉 2/3 杯、羅勒葉末 3 大匙、迷迭香末 2 大匙、百里香末 2 大匙、蒜末 2 大匙、檸檬皮末 1 大匙、柳丁皮末 1 大匙

調味料

粗粒黑胡椒粉適量、鹽適量

做法

1. 魚片先淋上白酒，再撒上醃料中的粗粒黑胡椒粉、鹽，放置 30 分鐘；四季豆切斜段；沾裹料全部拌勻。

2. 將魚片先沾裹麵粉，再充分地沾上蛋液，然後均勻地裹上拌勻的沾裹料。

3. 鍋中倒入 2 大匙油，待油熱後放入魚片煎至兩面都熟，取出。

4. 鍋中倒入 1/2 大匙油，待油熱後倒入四季豆炒熟，加入調味料，放入奶油、蒜末快炒幾下拌勻，盛入器皿中，鋪上魚片，也可擺上香草裝飾。

Use This Pot
Frying Pan

Ingredients
3 slices halibut, bream or salmon (approximately 600g), 100g green beans, 4T flour, 2 portions of egg liquid, 1/2T butter, 1/2T minced garlic, 2 1/2T cooking oil

Heading Format
2T white wine, coarsely ground black pepper and salt (proportion 2:1)

Coating
2/3C bread crumbs, 3T minced sweet basil, 2T minced rosemary, 2T minced thyme, 2T minced garlic, 1T minced lemon peel, 1T minced orange peel

Seasonings
coarsely ground black pepper and salt as needed

Methods
1. Drizzle fish slices with white wine, then sprinkle with coarsely ground black pepper and salt, let sit for 30 minutes until the flavor is absorbed. Cut green beans diagonally into sections. Combine the coating ingredients well.

2. Coat fish slices evenly with flour, dip in eggs completely, then coat evenly with the coating mixture.

3. Heat 2T of cooking oil in pan until smoking, fry fish until done on both sides, then remove.

4. Heat 1/2T of cooking oil in pan until smoking, then stir-fry green beans until cooked, season with seasonings to taste. Add butter and minced garlic, stir rapidly until well-mixed, then remove to a serving plate and spread fish on top. Garnish with herbs and serve.

CC 烹調祕訣 | Cooking tips

沾裹料中的新鮮香草末一定要充分擦乾才能使用。如果買不到新鮮的百里香，可用 1/2 大匙的乾燥百里香末取代。此外，臨時買不到羅勒的話，可以九層塔取代。

The fresh herbs in the coating ingredients have to be dried completely with a paper towel before use. If fresh thyme is not available, use 1/2T of dried thyme instead. If sweet basil is not available, substitute with basil.

地中海風味鮭魚

Mediterranean Style Salmon

在這道料理中，香濃的奶香佐以鮮嫩的鮭魚，令人垂涎三尺，而最後加入點檸檬皮末來提味，正是地中海料理的特色之一。

Fresh tender salmon with a thick creamy aroma. With a little touch of lemon peel, this mouth-watering dish is a signature dish of Mediterranean cuisine.

使用這個鍋一鍋到底
平底鍋

材料
去骨鮭魚 3 片（約 600 克）、西洋芹 1/2 支、牛蕃茄 1 個、黃甜椒 1/4 個、洋蔥末 1 大匙、蝦夷蔥末 1 大匙、檸檬皮末 1 大匙、白酒 80c.c.、鮮奶油 120c.c.、奶油 1/2 大匙

醃料
白酒 50c.c.、粗粒黑胡椒粉、鹽（量為 2：1）、茵陳蒿 1/3 大匙

調味料
粗粒黑胡椒粉適量、鹽適量

做法
1. 西洋芹削除外層粗纖維後切小丁；牛蕃茄、黃甜椒都切小丁。
2. 鮭魚洗淨後淋上醃料中的白酒，再撒上其他醃料，放置 30 分鐘。
3. 鍋燒熱，將鮭魚皮朝鍋面放入，煎至肉兩面都熟，取出。
4. 鍋中倒入奶油，待奶油融化後放入洋蔥末爆香，續入西洋芹炒香，再放入牛蕃茄、黃甜椒、調味料快炒幾下後起鍋。
5. 將白酒倒入做法 4. 的鍋中，煮至酒精揮發，倒入鮮奶油煮至濃稠狀，加入調味拌勻，再加入炒好的蔬菜料、蝦夷蔥末拌勻，淋在鮭魚上，撒些檸檬皮末即可。

Use This Pot
Frying Pan

Ingredients
3 pieces boneless salmon (approximately 600g), 1/2 minced celery stalk, 1 beef steak tomato, 1/4 yellow bell pepper, 1T minced onion, 1T minced chives, 1T minced lemon peel, 80c.c. white wine, 120c.c. whipping cream, 1/2T butter
Marinade:
50c.c. white wine, coarsely ground black pepper and salt (proportion 2:1), 1/3T tarragon

Seasonings
coarsely ground black pepper and salt as needed

Methods
1. Peel off the outer coarse layer of the celery, then dice celery stalk finely. Dice beef steak tomato and yellow bell pepper.
2. Rinse salmon and drizzle white wine from Marinade and sprinkle with remaining ingredients, then marinate for 30 minutes.
3. Heat pan and fry salmon with skin facing down until done on both sides, then remove from pan.
4. Melt butter in pan, stir-fry minced onion until fragrant, add minced celery to mix, then add beef steak tomato, yellow bell pepper and seasonings to taste. Saute rapidly for a minute and remove from heat.
5. Add white wine to the pan from method 4. cook until alcohol evaporates, pour in whipping cream and cook until thickened. Season with salt to taste, then return fried vegetables as well as chives to mix. Drizzle over salmon and sprinkle with minced lemon peel. Ready to serve.

CC 烹調祕訣 | Cooking tips
1. 因為鮭魚皮本身帶油，所以煎鮭魚時不需放油。
2. 檸檬皮不要削到皮白色的部分，不然味道會很苦
1. Since salmon skin contains oil, no oil is needed to fry the salmon.
2. When peeling the lemon peel, do not touch the white part, which tastes bitter.

歐風海鮮焗扇貝
European Cheese Scallops

喜歡吃海鮮的人可以換個新吃法，將海鮮料填在扇貝殼後撒入乳酪烹調，類似焗烤的口感，讓人回味再三！

People who love seafood can switch to a new cooking method. Stuff filling in a fan scallop shell, then sprinkle with cheese. It is similar to a baked dish, which offers a taste you will never forget!

使用這個鍋一鍋到底
平底鍋

材料

扇貝 8 ～ 10 個、蝦肉 80 克、白魚肉 80 克、牛蕃茄 45 克、西洋芹末 1 大匙、洋蔥末 1 大匙、白酒 60c.c.、帕瑪森乳酪 2 大匙、洋香菜末少許

白醬

奶油 1 大匙、麵粉 3 大匙、綜合高湯（做法參照 p.9）4 大匙、牛奶 5 大匙、乳酪絲 2 大匙、鹽適量

調味料

粗粒黑胡椒粉適量、鹽適量

做法

1. 扇貝淋上些許白酒，撒上少許粗粒黑胡椒粉；蝦肉、白魚肉都切丁；牛蕃茄切小丁。
2. 製作白醬：鍋中倒入奶油，待奶油融化後放入麵粉，一邊加入麵粉一邊攪拌，以小火炒約 8 分鐘，倒入高湯、牛奶拌勻至無顆粒，再加入乳酪絲、鹽拌勻成稠狀即可。
3. 將蝦肉、白魚肉、牛蕃茄、西洋芹末、洋蔥末放入盆中，倒入剩下的白酒，加入調味料拌勻，再倒入白醬拌勻。
4. 將做法 3. 填在扇貝上，撒上帕瑪森乳酪，排放在平底鍋中，蓋上鍋蓋，以小火烤至熟（也可移入烤箱中，以上下火 220℃ 烤熟），取出放在器皿中，撒些洋香菜末即可。

Use This Pot

Frying Pan

Ingredients

8～10 fan scallops, 80g shrimp meat, 80g white fish meat, 45g beef steak tomatoes, 1T minced western celery, 1T minced onion, 60c.c. white wine, 2T parmesan cheese, chopped parsley as desired

White Sauce

1T butter, 3T flour, 4T soup broth combo (see methods on p.9 for reference), 5T milk, 2T shredded cheese, salt as needed

Seasonings

coarsely ground black pepper and salt as needed

Methods

1. Drizzle fan scallops with a little white wine and sprinkle with a pinch of coarsely ground black pepper . Dice shrimp and white fish meat. Dice beef steak tomatoes.
2. To prepare White Sauce: Melt butter in pan, then fold in flour stirring and folding at the same time. Cook over low heat for about 8 minutes, then pour in soup broth and milk to mix. Stir until the lumps disappear, add cheese and salt to taste. Keep stirring until the sauce is thickened.
3. Place shrimp, white fish meat, tomatoes, minced celery and minced onion in a mixing bowl, pour in the remaining white wine, season with seasonings and mix well, then pour in white sauce and blend well.
4. Fill scallop shells with method 3., sprinkle with parmesan cheese, then line in order in frying pan and cover. Cook over low heat until done. (OR place in an oven and bake with upper and lower element at 220℃ until done). Transfer to a serving plate and sprinkle with parsley. Serve.

CC 烹調祕訣 | Cooking tips

由於使用平底鍋焗烤，所以不會像烤箱烤得那樣呈焦黃色，可撒上少許黑胡椒粉，比較像烤箱烤出來的顏色。

Because this dish is prepared in a frying pan, it does not appear as well-browned as would if it were cooked in the oven. Sprinkle with some coarsely ground black pepper to make it look more like an oven dish if preferred.

泰式紅咖哩焗烤生蠔
Thai Red Curry Oyster

平常不太敢吃生蠔的人，可以試試這道以紅咖哩烹調的泰式風味料理，顛覆你對生蠔西式吃法的既有想法。

People who dare not eat oysters can try this Thai style red curry dish. It might revolutionize your thinking about western oyster cooking.

使用這個鍋一鍋到底
平底鍋

材料
生蠔 8 個、香茅 1/2 支、檸檬葉 6 片、紅咖哩 1 大匙、小的雞蛋 1 個、椰漿 1/3 杯、紅辣椒末少許、香菜少許、油 1 大匙

調味料
魚露少許、細砂糖 2/3 小匙

做法
1. 生蠔洗淨，擦乾水分；香茅切段；取 2 片檸檬葉切末。
2. 鍋中倒入油，待油熱後放入紅咖哩，以小火炒至香味溢出，倒入椰漿煮至油浮出，放入香茅、4 片撕開的檸檬葉，以小火煮約 5 分鐘。
3. 先試一下味道，可斟酌加入少許魚露，倒入細砂糖拌勻，撈出香料，放至涼，然後加入雞蛋拌勻。
4. 將生蠔放入鍋中，淋入做法 3.，蓋上鍋蓋，烤至生蠔熟（也可移入烤箱中，以上下火 220℃烤至上色），取出撒些檸檬葉末、紅辣椒末，以香菜葉點綴即可。

Use This Pot
Frying Pan

Ingredients
8 uncooked oysters, 1/2 lemongrass stalk, 6 lemon leaves, 1T red curry, 1 small sized chicken egg,1/3C coconut milk, minced red chili pepper, cilantro as needed, 1T cooking oil

Seasonings
fish sauce as needed, 2/3t fine granulated sugar

Methods
1. Rinse oysters well, and dry with paper towel. Cut lemongrass stalk into sections. Mince two lemon leaves.
2. Heat oil in pan until smoking, stir-fry red curry over low heat until flavor is released, pour into coconut milk and cook until the oil floats on the surface. Add lemongrass and 4 pieces of lemon leaves. Cook over low heat for about 5 minutes.
3. Taste first to decide how much fish sauce needs to be added, then add granulated sugar. Remove and discard the herbs. Let cool and add egg to mix.
4. Remove oysters in pan and drizzle with method 3.. Cover and simmer until the oysters are done. (Or place in oven and bake with the upper and lower elements at 220℃ until brown.) Remove and sprinkle with minced lemon leaves and red chili pepper,sprinkle with cilantro Serve.

CC 烹調祕訣 | **Cooking tips**

也可使用貽貝來製作這道菜。此外，若不喜歡紅咖哩的辣味，可改用味道較溫和的黃咖哩。
Mussels can be used in place of the oysters. Use milder flavor yellow curry if the red curry is too spicy for you.

義大利
馬斯卡邦鮭魚
Italian Mascarpone Salmon

馬斯卡邦是義大利有名的軟質乳酪，常用來做甜點，其中以提拉米蘇最為知名。其實馬斯卡邦乳酪也很適合烹調料理，濃醇的奶香味和鮭魚非常搭配。

Mascarpone is a very famous Italian soft cheese, mostly used in making desserts. Tiramisu is one of the most popular desserts made from mascarpone cheese. In fact mascarpone cheese is quite suitable for preparing dishes. Its thick buttery flavor goes very well with salmon.

使用這個鍋一鍋到底
平底鍋

材料
鮭魚 600 克、蘆筍 100 克、馬斯卡邦乳酪 100 克、白酒 100c.c.、綜合高湯 3 大匙（做法參照 p.9）、鮮奶油 70 克、奶油 1/3 大匙

醃料
白酒 3 大匙、檸檬汁 1 1/2 大匙、粗粒黑胡椒粉、鹽（量為 2：1）

做法
1. 鮭魚切成 3 塊，先淋入醃料中的白酒和檸檬汁，再撒上粗粒黑胡椒粉、鹽，放置 15 分鐘。
2. 蘆筍放入滾水中汆燙熟，取出放入器皿中。
3. 鍋燒熱，倒入奶油，待融化後將鮭魚皮朝鍋面放入，煎至肉兩面都熟，取出放在蘆筍上。
4. 將材料中的白酒倒入鍋內煮至酒精揮發，倒入高湯、鮮奶油，放入馬斯卡邦乳酪拌勻，加入調味料，等煮沸後整個淋在鮭魚上，也可以用紅胡椒粒點綴。

Use This Pot
Frying Pan

Ingredients
600g salmon, 100g asparagus, 100g Mascarpone cheese, 100c.c. white wine, 3T soup broth combo (see methods on p.9 for reference), 70g whipping cream, 1/3T butter

Marinade
3T white wine, 1 1/2T lemon juice, coarsely ground black pepper and salt (proportion 2:1),

Methods
1. Cut salmon into 3 equal pieces, drizzle with white wine and lemon juice from Marinade, then sprinkle with coarsely ground black pepper and salt, let sit for 15 minutes until flavor is absorbed.
2. Blanch asparagus in boiling water until done, remove to a serving plate.
3. Heat frying pan and add butter until melt, then place salmon in pan with the skin facing down, cook until done both sides, then remove to the top of the asparagus.
4. Cook white wine from the Ingredients until the alcohol evaporates, pour in soup broth, add whipping cream and Mascarpone cheese to mix well. Bring to a boil and drizzle over salmon,sprinkle with red pepper corn.Serve.

CC 烹調祕訣 Cooking tips

除了鮭魚，亦可改用鱈魚來做這道菜。
In addition to salmon, cod may be used in this dish.

翡冷翠大蝦鑲鮮魚
Firenze Fish-Stuffed Prawn

這道3～4人份，品相佳且口感豐富的海鮮料理，是我的私房宴客菜之一。只要食材新鮮，簡單的平底鍋或炒鍋就能做出不輸大廚的豪華菜色。

This lavish and delicious seafood dish is one of my household specialties for entertaining guests. If the ingredients are fresh, even a simple frying pan or wo can prepare a table full of luxurious dishes.

使用這個鍋一鍋到底
平底鍋或炒鍋

材料

草蝦或明蝦 4 尾、去皮去骨白魚肉 120 克、洋蔥末 2 大匙、新鮮香菇末 3 大匙、蘑菇末 2 大匙、芹菜末 2 大匙、玉米粉 1 大匙、白酒 80c.c.（約 5 1/2 大匙）、洋香菜適量、油約 1 大匙

調味料

粗粒黑胡椒粉適量、鹽適量

醬汁

洋蔥末 1/2 大匙、蕃茄粒罐頭 1/4 罐、綜合高湯 1/4 杯（做法參照 p.9）、奧勒岡 1/2 小匙、月桂葉 1 片、油 1/4 大匙

做法

1. 蝦子由背部劃開，攤平取出腸泥，切斷蝦筋防止烹調過程中捲起，然後淋入些許（約 1 大匙）白酒，撒入調味料；白魚肉剁碎。

2. 鍋中倒入油，待油熱後放入洋蔥末爆香，續入香菇末、蘑菇末略炒幾下，再加入芹菜末炒幾下，加入調味料拌勻，放涼。

3. 魚肉放入盆中，倒入 1/2 大匙白酒、少許黑胡椒粉、鹽、玉米粉和做法 2. 拌勻成餡料，然後鋪在蝦子上。

4. 將做法 3. 放在平底鍋內，淋入 4 大匙白酒，蓋上鍋蓋，以小火烤至魚肉熟了（也可移入烤箱中，以上下火 220℃ 烤熟）。

5. 製作醬汁：鍋中倒入 1/4 大匙油，待油熱後放入洋蔥末爆香，再倒入其他所有醬汁中的材料煮約 6 分鐘，然後以調理機打成泥狀即可。

6. 將醬汁鋪在器皿中，放上蝦子，以洋香菜點綴即可。

Use This Pot
Frying Pan Or Wok

Ingredients

4 grass shrimps or prawn, 120g white fish meat with no skin and no bone, 2T minced onion, 3T fresh minced shiitake mushroom, 2T minced mushroom, 2 minced celery stem, 1T cornstarch, 80c.c. white wine (at least 5 1/2T), parsely as needed, 1T cooking oil

Seasonings

Coarsely ground black pepper as needed, salt as needed

Sauce

1/2 minced onion, 1/4 canned whole tomato, 1/4 soup broth combo (see methods on p.9 for reference), 1/2t oregano, 1 bay leaf, 1/4T cooking oil

Methods

1. Slice shrimp open from the back lengthwise, spread open and devein, cut off the tendons to prevent them from shrinking during cooking, then drizzle with a little (about 1 tablespoon) white wine and sprinkle with seasonings to taste. Crush white fish meat well.

2. Heat oil in pan until smoking, stir-fry minced onion until fragrant, add minced shiitake mushrooms and mushroom to mix. Saute for a minute, add minced celery and seasonings to taste. Remove from heat and let cool.

3. With fish meat in mixing bowl add 1/2T of white wine, along with coarsely ground black pepper, salt, cornstarch and method 2.. Stir until the mixture is evenly mixed, then spread evenly over shrimp.

4. Cook method 3. in frying pan, drizzle with 4T of white wine. Cover and simmer over low heat until the fish is cooked. (Or place in oven, bake with upper and lower element at 220℃ until done.)

5. To prepare Sauce: Heat 1/4T of oil in pan until smoking, stir-fry minced onion until fragrant, add the remaining ingredients and continue cooking for 6 more minutes, then remove to food processor and blend until mashed.

6. Spread sauce over the serving plate and top with stuffed shrimp, sprinkle with parsely. Serve.

CC 烹調祕訣 | Cooking tips

1. 蝦子本身已有鹹度，要注意不要撒入太多鹽，只要加入少量鹽來增加鮮甜度即可。
2. 魚肉煮至有彈性就代表熟了。

1. Shrimp is salty, so do not add too much salt. Use just a little to enhance the sweetness of the dish.
2. When the fish meat is firm, it means that it is done

南歐牛乾菌焗烤鱈魚

Southern European Porcini Mushrooms with Cod

牛肝菌是法國料理中的頂級食材，香氣濃郁，多用在燉飯、拌炒或製成醬汁。只要在料理中添加一點點，便能達到畫龍點睛的效果。

Porcini mushrooms with their strong aroma are a top ingredient of French cuisine. They are mostly used in risotto, stir-fries, or sauces. Just adding a little to the dish will raise the dish to the next level.

使用這個鍋一鍋到底
平底鍋

材料

去骨鱈魚 400 克、貽貝 8 個、乾燥牛肝菌 40 克、蘑菇 150 克、麵粉 3 大匙、麵粉適量（沾裹鱈魚用）、乳酪絲 3 大匙、蒜末 1/2 大匙、洋蔥末 3 大匙、牛奶 120c.c.、基本高湯適量（做法參照 p.6）、白酒 80c.c.、洋香菜少許、奶油 1 大匙、橄欖油 1/2 大匙、油 1 大匙

醃料

粗粒黑胡椒粉、鹽（量為 2：1）

調味料

粗粒黑胡椒粉適量、鹽適量

做法

1. 鱈魚擦乾後撒上醃料，放置 15 分鐘；牛肝菌放入溫水中浸泡 8 分鐘，取出瀝乾水分後切碎；蘑菇切片。

2. 鍋中倒入奶油，待奶油融化後慢慢放入麵粉，以小火炒至麵粉溢出香味，然後緩緩倒入牛奶、高湯拌勻，加入乳酪，加入鹽調味。

3. 另一鍋中倒入橄欖油，待油熱後放入蒜末、洋蔥末爆香，立即放入牛肝菌炒香，再放入蘑菇略炒幾下，倒入貽貝，淋入白酒煮至酒精揮發，加入調味料，然後倒入做法 2. 拌勻。

4. 鱈魚先沾裹麵粉。鍋中倒入油，待油熱後放入鱈魚煎至六分熟，取出放入烤皿中，淋入做法 3.，鋪上乳酪絲，放入鍋中，蓋上鍋蓋，以小火烤至乳酪絲融化（也可移入烤箱中，以上下火 220℃ 烤至呈金黃色），撒上洋香菜末即可。

Use This Pot
Frying pan

Ingredients

400g boneless cod, 8 mussels, 40g dried porcini mushroom, 150g mushrooms, 3T flour, flour as needed(for coating cod), 3T shredded cheese, 1/2T minced garlic, 3T minced onion,120c.c. milk, basic soup broth as needed (see methods on p.6 for reference), 80c.c. white wine, parsley as needed, 1T butter, 1/2T olive oil, 1T cooking oil

Marinade

coarsely ground black pepper and salt as needed (use golden proportion)

Seasonings

coarsely ground black pepper and salt (proportion 2:1)

Methods

1. Pat cod dry with paper towel, then sprinkle with marinade and let rest for 15 minutes. Soak Porcini mushrooms in lukewarm water for 8 minutes and remove to drain, then chop into small pieces. Cut mushrooms into slices.

2. Melt butter slowly in pan, then fold in flour a little at a time. Fry over low heat until the aroma of the flour is released, then gradually pour in milk and soup broth. Mix well and add cheese, then season with salt to taste.

3. Heat olive oil in pan until smoking, then stir-fry minced garlic and onion until fragrant, stir in Porcini mushrooms rapidly. Saute until fragrant, add mushrooms to mix, then add mussels. Drizzle with white wine and cook until alcohol evaporates, then season with seasonings to taste. Pour in Method 2. and mix well.

4. Dip cod in flour and coat evenly. Heat oil in pan until smoking, fry cod until 60% done, then remove to a baking bowl. Drizzle with Method 3. and spread shredded cheese evenly across it. Remove bowl to pan and cover. Cook over low heat until cheese melts. (Or place in oven and bake with upper and lower element at 220℃ until golden.) Sprinkle with minced parsley and serve.

CC 烹調祕訣 ▎Cooking tips

這道菜中可使用愛蒙特乳酪、莫札瑞拉乳酪或風提拿乳酪。
Emmental, mozzarella or fontina cheese is recommended in this dish.

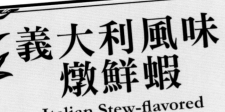

義大利風味燉鮮蝦

Italian Stew-flavored Fresh Shrimp

喜歡吃蝦子的人有福囉！這裡要介紹一道異國風味的蝦料理，材料中加入了蕈類、香草和鮮奶油燉煮，是有別於以往中式烹調的新吃法。

It is a fortunate day for those who love shrimp! We introduce here an exotic shrimp dish, in which ingredients such as mushrooms, herbs, and whipping cream are added. It is quite a new way of preparing shrimp, different from traditional Chinese methods.

使用這個鍋一鍋到底
燉鍋

材料

大尾蝦子 600 克、乾燥牛肝菌 30 克、新鮮
香菇 6 朵、蘑菇 10 朵、洋蔥末 1/4 杯、蒜
末 1 大匙、白酒 80c.c.、綜合高湯 1/2 杯（做
法參照 p.9）、鮮奶油 1/3 杯、月桂葉 1 片、
洋香菜末少許、油 1/2 大匙

調味料

粗粒黑胡椒粉適量、鹽適量

做法

1. 蝦子去殼留尾巴，由背部劃開，攤平取出
   腸泥；牛肝菌放入溫水中浸泡 10 分鐘，
   取出瀝乾水分後切絲；香菇、蘑菇切片。
2. 鍋中倒入油，待油熱後倒入洋蔥末、蒜末
   爆香，續入牛肝菌略炒幾下，再倒入香菇、
   蘑菇炒約 1 分鐘後加入蝦子，淋入白酒，
   待酒精揮發，倒入高湯燉煮，再倒入鮮奶
   油、月桂葉，加入調味料拌勻，燉煮至蝦
   子熟。
3. 將整鍋倒入器皿中，撒上洋香菜末即可
   食用。

Use This Pot
Stewpot

Ingredients

600g large shrimp, 30g porcini mushrooms, 6 fresh shiitake mushrooms, 10 button mushrooms, 1/4C minced mushroom, 1T minced garlic, 80c.c. white wine, 1/2C soup broth combo (see methods on p.9 for reference), 1/3C whipped cream, 1 bay leaf, minced parsley as needed, 1/2T cooking oil

Seasonings

coarsely ground black pepper as needed, salt as needed

Methods

1. Remove the shrimp shells and keep the tail, with the back sliced open lengthwise and devein. Add the porcini mushrooms to warm water and let soak for 10 minutes, then drain the water and shred the porcini mushrooms. Cut the shiitake mushrooms and mushroom into slices.
2. Heat oil into pan until smoking, stir-fry minced onion and minced garlic until fragrant. Add porcini mushrooms and shiitake mushrooms and mushroom, saute for 1 minute, then add shrimp. Drizzle with white wine and cook until the acohol evaporates, pour in the soup broth, then add whipping cream, bay leaf, and seasonings to taste. Stew until shrimp are done.
3. Pour the whole pot in a serving bowl and sprinkle with minced parsley. Serve.

CC 烹調祕訣 | Cooking tips

浸泡過牛肝菌的水不要倒掉，可以用來製作高湯，增加高湯的風味。
Do not throw away the water for soaking the porcini mushrooms. It can be used to prepare the soup broth to increase the flavor of the soup.

越南風味拌墨魚

Vietnamese Squid Cold Dish

這是一道東南亞風味的涼拌料理，清爽的醬汁搭配蔬菜和海鮮，是夏天最佳的開胃菜和下酒菜。

This is a Southeast Asian flavored cold dish. Its light clear dressing with vegetables and seafood is the best appetizer to go with beer on the hot summer days.

使用這個鍋一鍋到底
湯鍋或平底鍋

材料
墨魚（透抽）300 克、小黃瓜 1 條、胡蘿蔔
1/3 條、紅辣椒 2 條、西洋芹 60 克、熟的脆
花生適量

醬料
魚露 2 大匙、檸檬汁 3 大匙、細砂糖 1 大匙、
冷水 2 大匙、蒜末 2 大匙、紅辣椒末 1 大匙

做法
1. 墨魚放入滾水中汆燙，取出放涼，切成條狀；
   小黃瓜、胡蘿蔔和紅辣椒都切成絲；西洋芹
   切段。
2. 製作醬料：將所有材料拌勻。
3. 將做法 1. 放入盆中，淋入醬料拌勻，放置 6
   分鐘後盛入器皿中，撒上脆的熟花生碎即可。

Use This Pot
Soup Pot or Frying Pan

Ingredients
300g cuttlefish, 1 small cucumber, 1/3 carrot, 2
red chili peppers, 60g western celery, roasted
crispy peanuts as needed

Dressing
2T fish sauce, 3T lemon juice, 1T fine granulated
sugar, 2T cold water (cooked), 2T minced garlic,
1T minced red chili pepper

Methods
1. Blanch cuttlefish in boiling water for a second,
   remove to cool, then cut into strips. Shred
   Chinese cucumber, carrot, and red chili pepper
   finely. Cut celery into sections.
2. To prepare dressing: Combine all ingredients in
   bowl and mix well.
3. Place method **1.** in a mixing bowl, drizzle with
   dressing and mix well. Let sit for 6 minutes,
   then remove to a serving bowl, sprinkle with
   chopped peanuts and mint leaves. Serve.

CC 烹調祕訣 | Cooking tips

這道醬料可以運用在涼拌雞肉、越南風味涼拌青木瓜或者涼拌海鮮一起食用。
**This dressing can be used on cold chicken dishes, Vietnamese papaya salad, or any cold seafood
dishes.**

築地海鮮黃金捲

Fresh Golden Seafood Rolls

將雞肉混合各式海鮮做成餡料，加入荸薺更有咀嚼感。只要備好最新鮮的食材和乾淨的炸油，你一定會愛上這道清爽不膩的海鮮味。

The chicken is combined with many kinds of seafood to make the filling and water chestnut is added to enhance its chewy texture. You will fall in love with this clear light seafood flavor. Be sure to use fresh ingredients and clean frying oil.

使用這個鍋一鍋到底
平底鍋或炒鍋

材料
去皮去骨雞胸肉 200 克、去皮去骨白魚肉、蝦仁 120 克、透抽 100 克、山藥泥 3 大匙、荸薺碎 2 大匙、芹菜末 2 大匙、胡蘿蔔末 2 大匙、雞蛋 1 個、太白粉 1 1/2 大匙、腐衣 4 張

麵糊
低筋麵粉 120 克、蛋黃 2 個、冰水 160c.c.、鹽 1 小匙

調味料
鹽適量、白胡椒粉 1 小匙、香油 1 大匙

做法
1. 雞胸肉剁成泥；蝦仁、透抽都切丁。
2. 製作麵糊：將低筋麵粉過篩，然後和其他麵糊的材料拌勻。
3. 將雞胸肉、白魚肉、蝦仁、透抽、山藥泥、荸薺碎、芹菜末、胡蘿蔔末、雞蛋、太白粉放入調理盆，倒入調味料拌勻成有黏性的餡料。
4. 取出一張腐衣切成 2 份，攤開腐衣，鋪上餡料後捲成長筒狀，然後沾上麵糊。
5. 起油鍋，待油溫至 160℃，放入腐衣捲炸至呈金黃色，撈出瀝乾油分。

Use This Pot
Frying Pan Or Wok

Ingredients
200g skinless and boneless chicken breast, 120g skinless and boneless white fish meat, 120g shrimp meat, 100g squid, 3T Japanese mashed yam, 2T water chestnut, 2T minced celery, 2T minced carrot, 1 egg, 1 1/2T cornstarch, 4 tofu skin sheets

Batter
120g cake flour, 2 egg yolks, 160c.c. iced water, 1t salt

Seasonings
salt as needed. 1t white pepper, 1T sesame oil

Methods
1. Chop chicken breast finely until mashed. Dice shrimp and squid.
2. To prepare batter: Sift cake flour first, then mix well with the remaining ingredients.
3. Combine chicken breast, white fish meat, shrimp, squid, Japanese yam, water chestnut, minced celery, minced carrot, egg, cornstarch and seasonings well in a mixing bowl, keep stirring until the mixture is even and elastic.
4. Place a sheet of tofu skin on the working table, cut into two equally, spread the filling on one side of the sheet and roll up into a cylinder roll, then dip in batter evenly.
5. Heat cooking oil in frying pan until the temperature reaches 160℃, deep-fry the rolls until golden. Remove and drain. Serve.

CC 烹調祕訣 ▌Cooking tips

1. 材料中的荸薺必須擠乾水分，餡料才不會生水，影響口感。
2. 每取出一張腐衣操作時，必須將裝其他腐衣的袋子封好，避免腐衣因空氣變乾，也不可以碰到水，會發霉。
1. The water chestnut from the Ingredients must be squeezed dry, or it will keep releasing liquid and making the filling soggy.
2. Every time a tofu skin sheet is used, seal those remaining in the bag well to prevent them from drying out or becoming moldy due to contact with water.

薩丁尼亞蕃茄焗海鮮

Sardinia Style Braised Seafood

大部分的人對焗料理的接受度很高，我要介紹的這道異國風海鮮不僅美味、食材豐盛，而且品相極佳，招待朋友再適合不過。

Most people welcome braised dishes.
This foreign seafood dish with an
abundance of delicious ingredients is
not only tasty, but also looks good. It is a
perfect dish for entertaining friends.

使用這個鍋一鍋到底
平底鍋

材料

蛤蜊 200 克、蝦肉丁 4 大匙、透抽丁 4 大匙、蟹肉或蟹管肉丁 2 大匙、大的牛蕃茄 4 個、洋蔥末 1/2 大匙、西洋芹末 1 大匙、蘑菇丁 2 大匙、白酒 2 大匙、白醬 8 大匙（做法參照 p.75）、莫札瑞拉乳酪絲適量、油 1/2 大匙

調味料

粗粒黑胡椒粉適量、鹽適量

做法

1. 牛蕃茄往蒂頭下方約 0.8 公分切開，挖出籽後倒出汁液。
2. 將蛤蜊倒入鍋中，淋入 1 大匙白酒，煮至蛤蜊殼打開，取出肉。
3. 鍋中倒入油，待油熱後先放入洋蔥末爆香，續入西洋芹末炒香，再倒入蝦肉丁、透抽丁和蟹肉丁、蘑菇丁，淋入 1 大匙白酒略炒幾下，倒入調味料拌勻。
4. 將蛤蜊肉和做法 3. 倒入盆中，加入白醬、2 大匙乳酪絲拌勻成餡料。
5. 將餡料填入做法 1. 的牛蕃茄盅內，鋪上適量的乳酪絲，整個移入鍋內，蓋上鍋蓋，以小火烤至乳酪絲融化即可。

Use This Pot
Frying Pan

Ingredients

200g clams, 4T diced shrimp, 4T diced squid , 2T diced crab meat or crab leg meat , 4 large sized beef steak tomatoes, 1/2T minced onion, 1T minced celery, 2T wine wine, 8T white dressing (see methods on p. 75 for reference), 2T diced mushrooms, shredded mozzarella cheese as needed, 1/2T cooking oil

Seasonings

coarsely ground black pepper as needed, a pinch of salt

Methods

1. Cut the top of tomatoes off by removing the stem approximately 0.8cm from the stem, remove and discard seeds, and pour out the liquid.
2. Cook clams in pan with 1T white wine drizzled on top, cook until clams open and remove the meat.
3. Add some oil to pan and heat until steaming, stir-fry minced onion until fragrant, then add minced celery and fry until the flavor is released, add diced shrimp, squid ,crab and mushrooms meat. Drizzle with 1T of white wine and stir to mix. Season with seasonings to taste and mix well.
4. Remove clam meat and method 3. to a mixing bowl, add white sauce and 2T shredded mozzarella cheese. Mix well to be the filling.
5. Stuff the filling in the method 1. tomatoes, spread mozzarella evenly across top and remove the stuffed tomatoes to pan. Cover and bake over low heat until cheese melts. Remove and serve.

CC 烹調祕訣 ▌Cooking tips

1. 做法 5. 若想以烤箱操作的話，上下火大概用 230℃。
2. 西洋芹的外層纖維較粗，必須先刨掉外層再使用。

1. If you want to use an oven in method 5 , set upper and lower element to 230℃.
2. It is better to peel the outer coarse layer of the western celery before cooking.

利波諾
海鮮沙拉

Lipno Seafood Salad

這是為一家四口所設計的輕食料
理，豐盛的海鮮搭配各種翠綠的生
菜，享受口腹之外，更能吃得健
康、清爽。

This light dish is perfect for a family of four.
The rich seafood is filled with many kinds of
fresh vegetables. In addition to enjoying the
food, it is also healthy and light.

使用這個鍋一鍋到底
湯鍋或平底鍋

材料
透抽 200 克、新鮮鮑魚 4 個、蝦子 12 尾、洋蔥 1/2 個、牛蕃茄 1 個、各種生菜適量、西洋芹 1 支、白酒 2 大匙

沙拉醬
檸檬汁 2 大匙、柳丁汁 3 大匙、粗粒黑胡椒粉適量、鹽適量、橄欖油 300c.c.、白酒醋 30c.c.

做法
1. 透抽切圈狀；一半洋蔥切塊；牛蕃茄切片成船狀。將透抽放入鍋中，排上鮑魚後放上蝦子，再鋪放洋蔥塊、半支西洋芹，淋入白酒和 1 杯水，蓋上鍋蓋，煮至海鮮熟，取出放涼，蝦子去殼留尾巴。
2. 1/2 支西洋芹刨去外層較粗的纖維，然後切絲；生菜洗淨；另一半洋蔥切絲。將這些材料放入冰水中冰鎮約 15 分鐘，取出瀝乾水分。
3. 製作沙拉醬：將所有材料以打蛋器拌勻。
4. 將海鮮食材、西洋芹絲和洋蔥絲放入盆中，淋入 3 大匙沙拉醬拌勻，放置約 8 分鐘使其入味。
5. 將生菜擺在器皿中，放上做法 4. 和牛蕃茄，淋入沙拉醬即可食用。

這樣做更省時
Time-saving method

做法 1. 中煮海鮮時，若使用瑞康屋的平底鍋，只要倒入 2 大匙水，蓋上鍋蓋，待冒出煙改小火繼續加熱 2 分鐘即可。
When cooking method 1., use the Rikon Saute pan. Use only 2T of water, cover the pot, cook until smoke released, then reduce heat to low and continue cooking for 2 more minutes.

Use This Pot
Soup Pot Or Frying Pan

Ingredients
200g squid, 4 fresh abalone, 12 shrimp, 1/2 onion, 1 beefsteak tomato, a variety of fresh lettuce, 1 stem celery, 2T white wine

Salad dressing
2T lemon juice, 3T orange juice, coarsely ground black pepper as needed, salt as needed, 300c.c. olive oil, 30c.c white wine vinegar

Methods
1. Cut the squid into circles. Cut the onion into chunks. Cut beef steak tomato into a boat shape. Put the squid into the pot, with abalone on top of shrimp, then spread onion chunks and 1/2 celery across top, then drizzle in white wine and one cup of water. Cover and cook the seafood until it is done, then remove to cool. Discard the shrimp shell and keep the tail.
2. Peel off the outer coarse fiber from celery and discard, then shred celery finely. Wash the fresh lettuce and shred half of onion. Soak in ice water for about 15min, then drain the water.
3. To make salad dressing: Combine all the ingredients together and beat with an egg beater until evenly mixed.
4. Place seafood ingredients, celery stem, and shredded onion in a mixing bowl, drizzle with 3T of salad dressing, mix well and sit for 8 minutes until flavor is absorbed.
5. Place lettuce leaves in serving bowl, top with method 4. and beef steak tomato. Drizzle salad dressing over top. Serve.

CC 烹調秘訣 Cooking tips

使用材質佳的鍋具煮好的食材，味道鮮甜，吃得到食材原味
The dish is fresher and sweeter with a better quality cooking pan, which enables the real flavor of the food to be tasted.

里昂奶香焗海鮮
Lyon Seafood Combo

這是一道很適合在聚會或Party中享用的豐盛料理，不妨一次多製作幾盤，放在冰箱保存，隔天當成禮物再送給朋友，收到的人一定很開心！

This delicious dish is suitable for meetings and parties. If you wish to make large amounts, you can store it in the refrigerator, and give it to friends. The recipients will be very plea

使用這個鍋一鍋到底
湯鍋

材料

馬鈴薯 600 克、蝦仁 200 克、新鮮干貝 150
克、扇貝 8 個、透抽 120 克、鮮奶油 3 大匙、
帕瑪森乳酪粉 2 大匙、洋蔥末 2 大匙、芹菜
末 2 大匙、乳酪絲 160 克、麵包粉少許、洋
香菜末少許、白酒 80c.c.、奶油 1 1/2 大匙、
油 1 1/2 大匙

調味料

粗粒黑胡椒粉適量、鹽適量

做法

1. 馬鈴薯放入鍋中蒸爛，取出去皮搗成泥，
   加入 1 大匙奶油、鮮奶油、1 大匙帕瑪森
   乳酪粉和調味料拌勻；透抽切大丁。

2. 鍋中倒入 1/2 大匙奶油，待奶油融化後放
   入洋蔥末爆香，續入芹菜末炒香，倒入蝦
   仁、干貝、扇貝和透抽，淋入白酒煮至酒
   精揮發，加入調味料拌勻。

3. 將拌好的馬鈴薯泥、做法 2. 拌勻後倒入烤
   皿，鋪上乳酪絲，撒上剩下的帕瑪森乳酪
   粉和麵包粉，放入鍋中，蓋上鍋蓋，以小
   火烤至乳酪絲融化（也可移入烤箱中，以
   上下火 230℃烤至呈金黃色），撒上洋香
   菜末即可享用。

Use This Pot
Soup Pot

Ingredients
600g potatoes, 200g shelled shrimp, 150g scallops,
8 fan scallops, 120g squid, 3T whipping cream,
2T parmesan cheese powder, 2T minced onion,
2T minced celery,160g shredded cheese, flour as
needed, minced parsely as needed, 80c.c. white
wine, 1/2T butter, 1 1/2T cooking oil

Seasoning
coarsely ground black pepper and salt as needed

Methods

1. Steam potatoes in soup pot until soft, remove,
   discard skin and stir until mashed, add 1T butter
   and whipping cream, 1T Parmesan cheese
   powder, and ingredients from seasonings to mix.
   Chop squid into large chunks.

2. Pour 1/2T butter in pot and heat until it melts, add
   minced onion and cook until fragrant, fry minced
   celery until flavor is released, then put in shrimp,
   scallops, fan scallops, and squid. Drizzle with
   white wine and cook until alcohol evaporates.
   Add seasonings to taste and mix well.

3. Return mashed potatoes and ingredients from
   method 2. to mix, then pour into a baking bowl.
   Spread shredded cheese,the remaining Parmesan
   cheese, and flour evenly over top and place in a
   pot. Next, cover the pot, reduce heat to low and
   cook until cheese melts. (Or place baking bowl
   in oven and bake with upper or lower element at
   230℃ until golden). Sprinkle with minced parsely .
   Remove and serve.

CC 烹調秘訣 | **Cooking tips**

一般超市就能買到乳酪，可依個人喜好選用喜歡的口味。
Cheese can purchased at ordinary supermarket.Select any cheese as desired.

北歐風味海鮮

Northern European Style Stewed Seafood

朋友來家裡聚餐時，是否擔心不知該準備什麼？或者不想疲累做菜而無心招呼客人呢？以下這一鍋超豐盛的海鮮鍋可以讓你優雅地邊煮邊聊天，客人絕對吃得樂開懷。

Are you worried that you don't know what to prepare when friends come over? Or are you too tired to cook and reluctant to receive guests? This lavish seafood dish helps you cook elegantly yet make time for delightful conversation. Your guests will end up having a wonderful time.

使用這個鍋一鍋到底
燉鍋

材料

大蝦8尾、小蝦16尾、蛤蜊300克、扇貝4～
6個、貽貝8個、蒜末1大匙、洋蔥末3大匙、
蘑菇150克、蕃茄泥罐頭1/4杯、麵粉1大
匙、月桂葉1片、百里香2支、白酒150c.c.、
檸檬汁1～2個份量、綜合高湯11/2杯（做
法參照p.9）、奶油1大匙

調味料

粗粒黑胡椒粉適量、鹽適量

做法

1. 蝦子去殼留尾巴；所有海鮮料洗淨；麵粉
   與高湯拌勻成麵粉高湯。
2. 白酒倒入鍋中煮至酒精蒸發。
3. 鍋中倒入奶油，待奶油融化後放入洋蔥末
   炒香，放入蘑菇略炒幾下，加入蒜末、蕃
   茄泥、月桂葉、百里香、麵粉高湯拌勻，
   煮至濃稠，然後倒入做法2.的白酒再煮約
   10分鐘。
4. 依序加入蛤蜊、大蝦、扇貝、貽貝、小蝦，
   煮至海鮮熟了，放入檸檬汁，加入調味料
   拌勻，盛入器皿中，以百里香點綴即可。

Use This Pot
Stewpot

Ingredients
8 large shrimp tails, 16 small shrimp tails, 300g clams,4~6 fan scallops, 8 mussels, 1T minced garlic, 3T minced onion, 150g mushroom, 1/4C canned tomato paste, 1T flour, 1 bay leaf, 2 sprigs thyme, 150c.c. white wine, 1~2 lemon, 1 1/2C soup broth combo (see methods on p.9 for reference), 1T butter

Seasoning
coarsely ground black pepper and salt as needed

Seasoning
Salt as needed,1/2T chicken-flavored powder

Methods
1. Remove shells from shrimp and retain the tails. Rinse all seafood ingredients. Combine flour and soup broth to make flour broth.
2. Heat white wine in stew pot until the alcohol evaporates.
3. Heat butter in pot until it melts, fry minced onion until fragrant, add mushroom and saute for a minute, then add minced garlic, mashed tomato, bay leaf, thyme, flour broth to mix. Cook until the liquid is thickened, pour in method 2. white wine, and continue cooking for about 10 minutes.
4. Add the following ingredients in order: clams, shrimps, fan scallops, mussels, and small shrimp. Stew until all seafood is done. Add lemon juice, season with seasonings to taste. Transfer to a serving bowl. Sprinkle with sprigs thyme. Serve.

CC 烹調祕訣 | **Cooking tips**

蕃茄泥罐頭在一般超市即可購得，蕃茄泥已有鹹度，調味時需多加注意。

Tomato paste can be purchased from any supermarket. It is already salty, taste first before adding seasoning.

泰式辣醬炒大蝦

Thai Style Spicy Prawns

今年到泰國旅行時，堂哥特別帶我前往一家餐廳吃這道招牌料理。蝦子沾上獨特的泰式調味料令人印象深刻，推薦給喜歡重口味的讀者。

When I was on vacation in Thailand this year, a cousin of mine brought me to a diner that served a dish made in the following manner. The shrimp with its unique sauce really impressed me. This recipe is for readers who like strong-flavored food.

使用這個鍋一鍋到底
炒鍋

材料

大明蝦或大草蝦 5 尾、大蒜 6 粒、蔥 2 支、
芹菜 3 支、紅辣椒 2 根、香菜適量、油 2
大匙

調味料

泰國蝦醬膏 2 大匙、魚露 1 大匙、椰子糖
1/2 大匙、老抽 1/2 大匙、泰式高湯 2 大匙
（做法參照 p.9）

做法

1. 蝦子對切，挑除腸泥；大蒜切片；蔥、紅
   辣椒和芹菜都切段。
2. 鍋中倒入油，待油熱後放入蝦子快炒幾
   下，取出，鍋子先不要洗。
3. 將蒜片放入炒蝦的鍋中爆香，續入蔥爆
   香，然後倒入紅辣椒、蝦子和調味料快
   炒至蝦子熟。
4. 倒入芹菜拌抄幾下，盛入器皿內，撒上香
   菜即可。

Use This Pot
Wok

Ingredients
5 large prawns, 6 garlic cloves, 2 scallion, 3 celery stems, 2 red pepper, cilantro as needed, 2T cooking oil

Seasoning
2T Thai shrimp paste, 1T fish sauce, 1/2T coconut candy, 1/2T soy sauce, 2T Thai style soup broth (See methods on p.9 for reference)

Methods
1. Halve prawns into twos, devein and clean well. Cut garlic cloves into slices. Cut scallions, red chili pepper and celery stems into sections.
2. Heat oil in pan until smoking, stir-fry prawns rapidly for a second and remove. Retain the remaining oil in pan, do not wash the pan.
3. Stir-fry garlic slices in the pan until fragrance is released, then add scallion sections. Wait until the flavor is released, add red chili pepper, prawns and seasonings to taste. Stir rapidly until the prawns are done through.
4. Add celery sections to mix, then remove to a serving plate and sprinkle with chopped cilantro. Serve.

CC 烹調祕訣 | **Cooking tips**

1. 如果買不到椰子糖，可用細砂糖取代。
2. 蝦醬膏與蝦膏不同，味道類似 XO 醬。
1. If coconut candy is not available, substitute with granulated sugar.
2. Shrimp sauce paste is different from shrimp paste; its taste is similar to XO Sauce.

葡萄牙海鮮鍋

Portuguese Seafood Pot

這鍋加入了大量新鮮的海鮮，經過適當地烹調後，不僅味道鮮美，食材的精華更濃縮到湯汁中。可以用湯汁搭配麵包，保證你會愛上。

An abundance of fresh seafood is used in this recipe. After proper preparation, not only is the flavor of dish delicious, but the essence of the ingredients melt into the soup. It can be served with bread. I promise that you will fall in love with it!

使用這個鍋一鍋到底
湯鍋

材料

螃蟹 2 隻、蛤蜊 300 克、扇貝 4 個、貽貝 8 個、透抽 1 隻、大蝦 4 尾、小蝦 8 尾、蕃茄粒罐頭 1/2 罐、洋蔥 1/2 個、甜椒 1 個、蒜末 1 1/2 大匙、月桂葉 1 片、白酒 100c.c.、綜合高湯 1 1/2 杯（做法參照 p.9）、洋香菜末少許、奶油 1 大匙

調味料

粗粒黑胡椒粉適量、鹽適量、匈牙利紅椒粉 1/3 大匙

做法

1. 將螃蟹洗淨，去腮後每隻切成 4 大塊；蛤蜊放入鹽水中浸泡，使其吐沙；透抽切圈狀；蝦子由背部劃開，攤平取出腸泥；蕃茄粒搗碎；洋蔥和甜椒切絲。

2. 鍋中倒入奶油，待奶油融化後放入蒜末爆香，續入洋蔥炒至呈透明，再放入甜椒、月桂葉炒一下。

3. 倒入螃蟹略炒，再倒入蛤蜊拌炒、透抽，然後依序加入扇貝、貽貝、大蝦略炒，接著加入小蝦，淋入白酒煮至酒精揮發，倒入高湯和蕃茄粒、調味料燉煮至海鮮食材熟即可，上桌前再撒入洋香菜末即可享用。

Use This Pot
Soup Pot

Ingredients

2 crabs, 300g clams, 4 fan scallops, 8 mussels, 1 squid, 4 large shrimp, 8 small shrimp, 1/2C whole tomato, 1/2 onion, 1 bell pepper, 1 1/2T minced garlic, 1 bay leaf, 100c.c. white wine, 1 1/2C soup broth combo (see methods on p.9 for reference), chopped parsley as needed, 1T butter

Seasonings

coarsely ground black pepper and salt as needed, 1/3T Paprika powder Paprika powder

Methods

1. Rinse crabs well, discard gills from both sides and cut each into 4 equal pieces. Soak clams in salt water to make them spit out any impurities. Cut squid into rings. Slice shrimp open from the back lengthwise, devein and rinse well. Crush tomatoes. Shred onion and bell pepper.

2. Melt butter in pan until fragrant, stir-fry minced garlic until flavor is released, add onion and stir until transparent. Add bell pepper and bay leaf to mix.

3. Add crab and stir for a minute, then add clams and squid to mix. Next add fan scallops, mussels, large shrimp in order to mix, then small shrimp and drizzle with white wine. Cook until the wine evaporates, pour in soup broth ,tomato and add seasonings to taste. Cook until seafood is done, sprinkle with chopped parsley right before serving.

CC 烹調祕訣 | Cooking tips

這一道菜並非湯品，所以高湯不需倒入太多，可搭配麵包食用。

This dish is in fact not a soup dish, therefore there is no need to add too much soup broth. It can be served with bread.

日式炸蝦
蔬菜捲

通常油炸料理總讓人感到油膩，但若能搭配新鮮且口感清脆的生菜食用，真有說不出的清爽。

Japanese Deep-Fried Shrimp Veggie Rolls

Deep-fried dishes are usually very oily. The fresh and crunchy lettuce reduces the oily feel and makes the dish much lighter.

使用這個鍋一鍋到底
平底鍋

材料
蝦子 4 尾、羅蔓生菜 5 片、麵粉適量

沙拉醬
美乃滋 50 克、檸檬汁 1/2 大匙、山葵 1 小匙

麵衣
冰過的蛋黃 1 個、冰過的低筋麵粉 80 克、
冰水 140c.c.、鹽 1 小匙、細砂糖 1/2 小匙

做法
1. 生菜洗淨後放入冰水中冰鎮約 15 分鐘，
   取出瀝乾水分；蝦子由背部劃開，攤平取
   出腸泥，切斷蝦筋防止烹調過程中捲起。
2. 製作麵衣：先將蛋黃和冰水拌勻，再加入
   麵粉、鹽和細砂糖拌勻。
3. 蝦子依序沾裹麵粉、麵衣。起油鍋，待油
   溫至 160℃，放入蝦子炸至呈金黃色，撈
   出瀝乾油分。
4. 製作沙拉醬：將所有材料拌勻後裝在擠
   花袋中。
5. 取一片生菜切成細絲。將一片生菜放在盤
   中，鋪上生菜絲，擠些沙拉醬，排放蝦子，
   最後再擠些沙拉醬即可。

Use This Pot
Frying pan

Ingredients
4 shrimp, 5 romaine lettuce leaves, flour as needed

Salad Dressing
50g mayonnaise, 1/2T lemon juice, 1t wasabi paste

Batter Coating
1 refrigerated egg yolk, 80g refrigerated cake flour, 140c.c. iced water, 1t salt, 1/2t fine granulated sugar

Methods
1. Rinse lettuce well, let sit in ice water for about 15 minutes, then remove and drain. Slice shrimp open lengthwise from the back, spread out and devein, then break the tendon to prevent it from shrinking during cooking.
2. To prepare coating: combine egg yolks and ice water well, add flour, salt, and granulated sugar to mix.
3. Coat shrimp with flour first, then coat with batter evenly. Heat oil in pan until the temperature reaches 160℃, deep-fry shrimp until golden and place in drain.
4. To prepare salad dressing: combine all the ingredients well and place in a pastry bag.
5. Shred a lettuce leaf finely. Line a lettuce leaf on a serving plate, place shredded lettuce on top, squeeze salad dressing over the lettuce, then arrange shrimp on top and squeeze more salad dressing on the shrimp. Ready to serve.

CC 烹調祕訣 | Cooking tips

1. 使用冰過的蛋黃、麵粉和冰水來製作麵衣，可使炸好的麵衣更酥脆。
2. 如果沒有擠花袋，可將沙拉醬裝在塑膠袋中，在角落剪一個小洞使用即可。
1. Using refrigerated egg yolks, flour, and ice water to prepare the flour batter coating will make the coating more crispy after deep-frying.
2. If a pastry bag is not available, hold the salad dressing in a plastic bag and cut a small hole at the corner.

貝尼特海鮮墨魚義大利麵

Benet Seafood with Squid Ink Linguine

這道墨魚口味可以說是最受歡迎的義大利麵前三名，而我自己則喜歡在墨魚之外，再搭配其他海鮮料，吃起來更過癮。

This cuttlefish flavored linguine is one of the top three pasta dishes. I myself like to add other seafood in addition to the cuttlefish, which creates more satisfaction.

使用這個鍋一鍋到底
平底鍋

材料

蝦子 12 尾、蛤蜊 200 克、小墨魚 16 隻或中墨魚 4 隻、墨魚圈 150 克、牛蕃茄 2 個、墨魚麵 1 包、白酒 150c.c.、蒜末 3 大匙、紅辣椒末 1 1/2 大匙、墨魚醬 1 大匙、帕瑪森乳酪粉適量、洋香菜末少許、橄欖油 1 1/2 大匙

煮麵料

滾水適量、鹽 1 1/2 大匙、橄欖油約 2 大匙

調味料

粗粒黑胡椒粉適量、鹽適量

做法

1. 蝦子去殼留尾巴，挑除腸泥；蛤蜊放入鹽水中浸泡，使其吐沙；牛蕃茄去皮後切丁。
2. 準備煮麵。將墨魚麵放入滾水中，加入鹽、橄欖油，煮約 10 分鐘後撈起，瀝乾水分（依品牌不同，煮的時間可參照包裝袋上寫的時間）。
3. 小墨魚整隻塗抹橄欖油，放入鍋中煎熟，淋入 60c.c. 白酒煮至酒精揮發，取出。
4. 鍋中倒入橄欖油，待油熱後放入蒜末爆香至呈微黃色，倒入紅辣椒末炒香，加入蛤蜊煮至殼略開，再加入蝦子、小墨魚、墨魚圈，淋入剩下的白酒煮至酒精揮發，然後倒入墨魚醬、墨魚麵、牛蕃茄，加入調味料拌勻，盛入器皿內，撒上帕瑪森乳酪粉、洋香菜末即可。

Use This Pot

Frying pan

Ingredients

12 shrimp, 200g clams, 16 small sized cuttlefish, or 4 medium sized cuttlefish, 150g cuttlefish circles, 2 beef steak tomatoes, 1 pack squid ink linguine, 150c.c. white wine, 3T minced garlic, 1 1/2T minced red chili pepper, 1T squid ink sauce, parmesan cheese powder as needed, chopped parsley as needed, 1 1/2T olive oil

Ingredients for preparing linguine

Boiled water as needed, 1 1/2T salt, 2T (approximately) olive oil

Seasonings

coarsely ground black pepper and salt as needed

Methods

1. Remove shell from shrimp, retain tails and devein. Soak clams in salt water to let them spit out any impurities. Remove and discard the skin from beef steak tomatoes, then dice.
2. To prepare linguine: Cook linguine in boiling water with salt and olive oil for about 10 minutes, then drain. (Cooking time may differ by linguine brand. Reference the package directions.)
3. Coat whole small cuttlefish with olive oil, then fry in pan until done. Drizzle with 60c.c. white wine and cook until the alcohol evaporates, then remove from pan.
4. Heat 1 1/2T of olive oil in pan until smoking, stir-fry minced garlic until lightly brown, add minced chili pepper to mix. Cook clams until the shells are almost wide open, add shrimp, small cuttlefish, cuttlefish circle and drizzle with remaining white wine. Cook for a minute until the alcohol evaporates, season with squid ink sauce, return linguine, then add beef steak tomatoes. Season with seasonings to taste. Transfer to a serving plate and sprinkle with parmesan cheese powder and parsley. Serve.

CC 烹調祕訣 | Cooking tips

做法 3 中如果煮好的麵料太乾，可加入適量煮麵水或者高湯。

Add a little boiling water from cooking linguine or soup broth to method (3) if it turns out to be too dry.

蒙迪那炒野菇與西西里島海鮮湯，
想像義大利的熱情陽光；
和風四季豆搭配秋葵豆腐，
彷彿置身京都的寂靜中；
田園拌乾蕃茄佐以鮮蝦蟹肉南瓜湯，
體驗法國鄉村的樸實風味；

現在就翻開食譜，
和我一塊來趟餐桌上的旅行吧！

Imagine Modena Balsamico Mushroom and Sicily Seafood Soup as the passionate Italian sun. Eating Japanese Green Beans with Sesame Seeds and Kyoto Okra Tofu is like standing in a quiet Kyoto street. Feel the rich earth of the French countryside through Veneto Country Dried Tomato and Mushroom and French Style Pumpkin with Seafood Soup.
Let's open the cookbook and turn our dinner table into a journey to an exotic location.

PART 3
蔬菜和湯類
VEGETABLES AND SOUP SECTION

蒙迪那炒野菇
Modena Balsamic Mushrooms

位於義大利的蒙迪那（Modena）因盛產巴薩米克醋而聞名於世，這道利用巴薩米克醋烹調的菇類料理，帶有芬芳的酸甜風味，適合當作前菜或開胃菜享用。

Modena in Italy is world famous for its balsamic vinegar. This mushroom dish cooked with balsamic vinegar tastes a bit sour and sweet and is perfect as an appetizer or refreshment.

使用這個鍋一鍋到底
炒鍋或平底鍋

材料
蘑菇 200 克、新鮮香菇 200 克、培根 3 片、蒜末 2 大匙、紅辣椒末適量、巴薩米克醋 4 大匙、洋香菜少許、羅蔓生菜適量

調味料
粗粒黑胡椒粉 1 小匙、鹽適量

做法
1. 蘑菇切片；香菇切條；培根切絲。
2. 鍋中不放油，放入培根炒至酥脆，加入 1 大匙蒜末、紅辣椒末爆香，淋入巴薩米克醋，煮至濃縮剩 2 大匙，立即倒入蘑菇、香菇快炒到熟，加入剩下的蒜末、調味料拌勻。
3. 先將羅蔓生菜排入器皿中，再將做法 2. 盛入器皿中，撒些洋香菜即可。

Use This Pot
Wok Or Frying Pan

Ingredients
200g mushroom, 200g fresh shiitake mushrooms, 3 slices bacon, 2T minced garlic, red chili pepper as needed, 4T balsamic vinegar, parsley and romaine lettuce leaves as needed

Seasonings
1t coarsely ground black pepper, salt as needed

Methods
1. Cut mushroom into pieces. Cut shiitake into strips. Shred bacon.
2. Do not add oil in frying pan. Fry bacon in pan until crispy, add 1T minced garlic and red chili pepper until fragrant. Drizzle with balsamic vinegar and cook until about 2 tablespoons of the liquid is left. Add mushroom and shiitake mushrooms immediately, fry until done, then add remaining minced garlic and ingredients from seasonings to mix.
3. Line the serving plate with romaine lettuce leaves,then top with method 2., sprinkle with parsley. Serve.

CC 烹調祕訣 | Cooking tips

1. 蘑菇切開後容易氧化變黑，可拌入少許檸檬汁，以防止變黑。
2. 如果選用的是 10 年以上的巴薩米克醋，可不必加熱濃縮，直接將材料改成 2 大匙即可。
1. The mushroom darkens easily because of oxidation after being sliced. Add a little lemon juice to prevent that from happening.
2. If a 10 year old balsamic vinegar is used, it is not necessary to thicken the liquid. Just change the balsamic vinegar in the ingredients from 4T to 2T.

維內多田園拌乾蕃茄
Veneto Country Dried Tomato and Mushroom

這道料理最特別之處在於醬料！
微酸且層次豐富的口感令許多第
一次品嘗的人驚豔不已，忍不住
推薦給所有讀者。

The special part of this dish is its sauce.
It is slightly sour, yet has a rich textured
layer that always amazes people who
try it for the first time. I cannot help but
recommend it to all the readers.

使用這個鍋一鍋到底
平底鍋

材料
荷蘭豆 100 克、杏鮑菇 100 克、新鮮香菇 100 克、橄欖油適量（拌香菇用）、橄欖油約 3 大匙、帕瑪森乳酪粉 2 大匙、蝦夷蔥末 2 大匙

調味料
粗粒黑胡椒粉適量、鹽適量

醬料
蒜末 2 大匙、紅辣椒末適量、西洋芹末 2 大匙、乾蕃茄干末 2 1/2 大匙、紫蘇梅肉末 2 大匙、橄欖油 1 大匙

做法
1. 將杏鮑菇、香菇放入鍋中，倒入 1 杯水，以小火煮熟，取出切成條狀後放入盆中，倒入適量的橄欖油、調味料拌勻。
2. 荷蘭豆放入滾水中汆燙熟，取出切成細條狀。
3. 製作醬料：鍋中倒入橄欖油，待油熱後放蒜末炒香至呈金黃色，倒入紅辣椒末、西洋芹末炒勻，再放入乾蕃茄干末、紫蘇梅肉末炒勻，加入調味料拌勻。
4. 將做法 1. 和做法 2. 拌勻後倒入器皿中，淋入醬料，最後撒入橄欖油、帕瑪森乳酪粉、蝦夷蔥末即可。

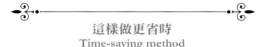

這樣做更省時
Time-saving method

做法 1. 中煮菇類時，若使用瑞康屋的平底鍋，只要倒入 3 大匙水，蓋上鍋蓋，待冒出煙改小火繼續加熱 3 分鐘即可取出。

In method **3.** if a saute pan is used to cook the mushrooms, only 3T of water is needed. Cook with top covered until smoke is released, reduce heat to low, and continue cooking for 3 minutes, then remove.

Use This Pot
Frying Pan

Ingredients
100g snow peas, 100g king oyster mushrooms, 100g fresh shiitake mushrooms, olive oil as needed, 3T olive oil, 2T parmesan cheese powder, 2T chive

Seasonings
coarsely ground black pepper and salt as needed

Sauce
2T minced garlic, red chili pepper as needed, 2T minced celery, 2 1/2T dried minced tomato, 2T minced perilla plum flesh, 1T olive oil

Methods
1. Cook king oyster mushrooms and shiitake mushrooms over low heat in pan with 1 cup of water until done. Remove and cut into strips, then remove to mixing bowl. Add olive oil, coarsely ground black pepper , and salt to taste, then mix well.
2. Blanch snow peas in boiling water until done, remove and cut into strips.
3. To prepare sauce: Heat olive oil in pan until smoking, stir-fry minced garlic until golden and fragrant, add chili pepper and celery to mix. Add dried tomato and perilla plums, mix well and add seasoning to taste.
4. Combine methods **1.** and **2.** in bowl, drizzle with sauce and olive oil. Sprinkle with parmesan cheese powder and minced chives over top. Serve.

CC 烹調祕訣 ▍Cooking tips

1. 進口的乾蕃茄干可以在販售進口食材的店，或者百貨公司食品櫃有售，包裝多為罐裝或零售。
2. 蝦夷蔥又叫細香蔥，細長管狀，多切成細末撒在料理上，用來調味。
1. Imported dried tomato can be purchased at the imported goods stores, or in the department store. It is mostly available in cans or retail packages.
2. Chives are fine, long, and tube-like. It is generally minced and sprinkled over the ingredients to season the dish.

義大利風味乳酪焗茄子

Italian Braised Eggplant with Cheese

蕃茄、羅勒、乳酪的絕妙組合，讓義大利料理激盪出令人讚嘆的美味。經典的蕃茄醬汁更能搭配其他食材，妙用多多。

The combination of tomato, basil, and cheese brings out the fantastic aroma of Italian cuisine. Classic tomato sauce, so useful, can be paired with many other ingredients.

使用這個鍋一鍋到底
平底鍋

材料
茄子 300 克、莫札瑞拉乳酪 650 克、羅勒
適量、洋香菜末適量、麵粉適量、奶油 1/2
大匙、帕瑪森乳酪粉 1 大匙、炸油適量

調味料
粗粒黑胡椒粉、鹽（量為 2：1）

蕃茄醬汁
蕃茄粒罐頭 1 罐、蒜末 1 1/2 大匙、洋蔥末
3 大匙、月桂葉 1 片、奧勒岡 1 小匙、基本
高湯 1/4 杯（做法參照 p.6）、油 1/2 大匙

做法
1. 茄子切片後撒上調味料，放置 6 分鐘，沾
   裹麵粉。起油鍋，待油溫至 160℃，放入
   茄子炸熟，撈出瀝乾油分。莫札瑞拉乳酪
   切片。
2. 製作蕃茄醬汁：蕃茄粒先切碎。鍋中倒入
   油，待油熱後放入蒜末爆香，再放入洋蔥
   末炒香且呈透明，倒入蕃茄粒和高湯、月
   桂葉、奧勒岡煮約 25 分鐘。
3. 鍋中抹上奶油，依序鋪上炸好的茄子，淋
   上蕃茄醬汁，鋪上羅勒，再鋪一層莫札瑞
   拉乳酪，這樣算一層，然後依照這個順序
   共鋪好三層。
4. 撒上帕瑪森乳酪粉，蓋上鍋蓋，以小火加
   熱至莫札瑞拉乳酪融化即可。

這樣做更省時
Time-saving method

製作蕃茄醬汁時，如果使用瑞康屋的平底
鍋，只要 12 分鐘就可以煮好了。
Using a saute pan to prepare the tomato sauce
takes only 12 minutes.

Use This Pot
Frying Pan

Ingredients
300g eggplants, 650g mozzarella cheese, sweet
basil and minced parsley as needed, flour as
needed, 1/2T butter, 1T ground parmesan cheese,
deep-frying oil as needed

Seasonings
coarsely ground black pepper and salt (proportion
2:1)

Tomato Sauce
1 can whole tomato, 1 1/2T minced garlic, 3T
minced onion, 1 bay leaf, 1t oregano, 1/4C basic
soup broth (see methods on p. 6 for reference) ,
1/2T cooking oil

Methods
1. Slice eggplant and sprinkle with seasonings, let
   sit for 6 minutes and coat evenly with flour. Heat
   oil in pan until the temperature reaches 160℃,
   deep-fry eggplant slices until done. Remove and
   drain well. Cut mozzarella cheese into slices.
2. To prepare tomato sauce: Cut whole tomatoes
   into pieces. Heat cooking oil in pan until smoking,
   stir-fry minced garlic until fragrant, add minced
   onion and fry until transparent, then add tomato
   as well as bay leaf, and oregano. Cook for
   approximately 25 minutes.
3. Grease the pan with butter, arrange eggplant
   slices at the bottom as the first layer, then drizzle
   tomato sauce over the eggplants. Spread sweet
   basil evenly across the eggplant slices as the
   second layer and then sprinkle mozzarella cheese
   on top to finish the last layer.
4. Sprinkle with ground parmesan cheese and
   cover, cook over low heat until the mozzarella
   cheese melts. Remove and serve.

CC 烹調祕訣 ▎Cooking tips

這道料理中的蕃茄醬汁，也可以用海鮮義大利麵中的炒紅醬取代。

The Red Sauce from Seafood Spagetti may be used in place of the tomato sauce in this recipe.

和風四季豆
Japanese Green Beans with Sesame Seeds

冷熱皆宜的一道日式小菜，搭配冰涼的啤酒一塊享用，絕對令人滿足，而且成功率百分之百喔！

This Japanese can be served as either cold or hot. Enjoying it with beer will definitely satisfy your appetite. Its success rate is 100%

使用這個鍋一鍋到底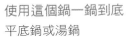
平底鍋或湯鍋

材料
四季豆300克、熟白芝麻少許

煮料
醬油5大匙、細砂糖2大匙、味醂2 1/2 大匙、
水 1/2 杯、柴魚精 1/2 大匙

做法
1. 四季豆放入鍋內，倒入煮料，以中小火煮
滾，然後改小火煮至四季豆變軟，撈出。
2. 將煮好的四季豆盛入器皿中，撒上熟白芝
麻即可。

Use This Pot
Frying Pan Or Soup Pot

Ingredients
300g string beans, roasted white sesame seeds as desired

Seasonings
5T soy sauce, 2T fine granulated sugar, 2 1/2T mirin, 1/2C water, 1/2T bonito extract

Methods
1. Place string beans in pan with all the seasonings added. Bring to boil over medium heat, then reduce heat to low and continue cooking until string beans are soft, then remove from pan.
2. Place string beans in serving bowl and sprinkle with white sesame seeds. Serve.

CC 烹調祕訣 ▌Cooking tips

如果不喜歡用柴魚精，可製作日式柴魚高湯來取代。先將一塊昆布放入鍋中，倒入 1/5 杯水加熱，
待煮沸後撈出昆布，倒入 1/5 杯柴魚片，約 5 秒後立即撈出柴魚片，煮太久的話湯汁腥味會太重。
接著將柴魚高湯與四季豆、醬油、細砂糖和味醂一起煮至軟即可。

If bonito extract is not preferred, substitute with Japanese style bonito soup broth. Cook a piece
of Kombu kelp in pan with 1/2C of water added, bring to a boil and remove the kombu. Pour in
1/5C of bonito flakes for approximately 5 seconds and remove. Cooking too long will make the
soup too fishy. Next combine the bonito soup broth with string beans, soy sauce, granulated sugar
and mirin. Cook until soft.

義式香果鹽炸蔬菜
Italian Deep-Fried Veggies with Fruit

這道料理是以檸檬皮、柳丁皮、粗粒黑胡椒粉、鹽混合而成的香果鹽來調味。水果清新的氣味、搭配炸蔬菜的清爽口感，深受大家喜愛，更推薦給素食讀者享用。

This dish is seasoned with salt combined with lemon zest, orange peel, and black pepper. The clear light fruit flavored with deep-fried vegetables is very popular among vegetarians. It is highly recommended.

使用這個鍋一鍋到底
平底鍋或炒鍋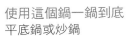

材料
南瓜 200 克、杏鮑菇 2 支、茄子 1 條、蘆筍 120 克、低筋麵粉適量、蛋液 2 個份量、麵包粉飾量、炸油適量

香果鹽
鹽 1 大匙、粗粒黑胡椒粉 1/2 大匙、柳丁皮末 1/3 大匙、檸檬皮末 1/3 大匙

做法
1. 南瓜去籽切片;杏鮑菇、茄子切片;蘆筍切約 10 公分長。
2. 製作香果鹽:將香果鹽的材料混合均勻。
3. 將所有蔬菜依序沾裹麵粉→蛋液→麵包粉。
4. 起油鍋,待油溫至 160℃,放入蔬菜炸至呈金黃色,取出瀝乾油分,盛入器皿中。食用時,沾上香果鹽即可享用。

Use This Pot
Frying Pan Or Wok

Ingredients
200g pumpkin, 2 king oyster mushrooms, 1 eggplant, 120g asparagus, cake flour as needed, 2 eggs portion, bread crumbs and deep-frying oil as needed

Fruit Salt
1T salt, 1/2T coarsely ground black pepper , 1/2T minced orange peel, 1/2T minced lemon peel

Methods
1. Remove pumpkin seeds from pumpkin and cut into slices. Cut king oyster mushrooms and egg-plant into slices. Cut asparagus into 10cm long sections.
2. To prepare fruit salt: Combine all the Fruit Salt in-gredients together.
3. Coat all vegetables and mushrooms with flour, then dip in eggs and then coat evenly with bread crumbs.
4. Heat oil in pan until the temperature reaches 160℃, deep-fry all the vegetables in oil until gold-en. Remove and drain, then line in serving plate. Serve fruit salt on the side as a dip.

CC 烹調祕訣 | Cooking tips

也可以把南瓜片或地瓜片沾裹麵衣炸來吃,同樣美味喔!

Pumpkin or yam can be sliced and coated with coating, then deep-fried. It's so delicious.

京都秋葵豆腐
Kyoto Okra Tofu

如同懷石料理般的精緻、口味清淡，是一道健康美味的和風開胃菜。加上做法簡單，料理新手絕對不失敗。

This is as exquisite as kaiseki - ryori cuisine, a light and healthy appetizer. It is very simple and easy to prepare. A first time cook will never fail.

使用這個鍋一鍋到底
平底鍋

Use This Pot
Frying Pan

材料
嫩豆腐 2 塊、秋葵 8 支、柴魚片適量

Ingredients
2 squares tender tofu, 8 okra, bonito flakes as needed

醬料
白味噌 2 大匙、淡色醬油 1 大匙、柴魚醬油 1 大匙、白芝麻醬 1 1/2 大匙、味醂 1 大匙、香油 1/2 大匙、細砂糖 1/3 大匙、日式高湯 3 大匙（做法參照 p.8）

Sauce
2T white miso paste, 1T light soy sauce, 1T bonito flavored soy sauce, 1 1/2T white sesame paste, 1T mirin, 1/2T sesame oil, 1/3T fine granulated sugar, 3T Japanese Soup Broth (see methods on p.8 for reference)

做法
1. 秋葵放入滾水中汆燙熟，取出切片；嫩豆腐也放入滾水中汆燙一下，取出瀝乾水分。
2. 製作醬料：將所有材料拌勻即可。
3. 將嫩豆腐放在器皿中，淋上醬料，鋪上秋葵，撒些柴魚片即可。

Methods
1. Blanch okra in boiling water until done, then remove and cut into slices. Blanch tender tofu in boiling water for a minute and remove to drain.
2. To prepare sauce: Combine all ingredients and mix well.
3. Place tender tofu in a serving plate and drizzle with sauce. Line okra on top and sprinkle with bonito flakes. Serve.

plus+
同場加映

涼拌秋葵 Okra Salad

材料
秋葵 150 克、醬料和柴魚片適量

Ingredients
150g okra, sauce and bonito flakes as needed

做法
將秋葵放入滾水中汆燙，取出瀝乾水分放在器皿中，淋上做法 2. 中的醬料，再撒些柴魚片即可。

Methods
Blanch okra in boiling water until done, then remove to drain, transfer to a serving plate. Drizzle with methods 2. sauce, sprinkle with bonito flakes. Serve.

CC 烹調祕訣 | Cooking tips

使用 Hotpan 烹調的話，只要加入 3 大匙水，就能將豆腐和秋葵汆燙熟。
If this dish is cooked with a hotpan, only 3T of water is needed to blanch the tofu and okra until done.

泰式涼拌什錦蔬果

Thai Style Cold Veggie and Fruit Mix

這道泰國常見的家常菜，是曼谷好友特別推薦的，加入了大量蔬果，並佐以獨特的泰式拌醬，更顯清爽可口，令人百吃不膩。

This home-style dish is very common in Thailand. It is highly recommended by a good friend from Bangkok. With its vegetables and fruits and unique Thai style dressing it is light and delicious. You will never get tired of eating it.

使用這個鍋一鍋到底
湯鍋或平底鍋

材料

蝦子 200 克、貽貝 8 個、蘋果 1 個、小蕃茄 8 個、青木瓜 100 克、紅辣椒 1 條、熟的脆花生 2 大匙、香菜少許

醬汁

酸子醬（羅望果）1 1/2 大匙、檸檬汁 3 大匙、魚露 3 大匙、椰子糖（棕櫚糖）1 1/2 大匙、蒜末 2 大匙

醃料

香茅末 1 大匙、蒜末 1 大匙、檸檬汁 1/3 大匙、魚露 1 小匙、細砂糖 1 小匙、九層塔末 1 大匙

做法

1. 蝦子去殼留尾巴，由背部劃開，攤平取出腸泥，放入滾水中氽燙熟，取出瀝乾水分。貽貝也氽燙熟，取出瀝乾水分。
2. 將蝦子、貽貝拌入醃料中，放置 15 分鐘。
3. 青木瓜削除外皮，挖掉籽後切絲；蘋果去籽後切片，浸泡鹽水 5 分鐘，取出瀝乾水分；小蕃茄每顆對切；紅辣椒切絲；花生搗碎。
4. 將醬汁的材料拌勻，然後把青木瓜、蘋果、小蕃茄拌入醬汁拌勻，放置 8 分鐘，再加入蝦子、貽貝、紅辣椒中拌勻，盛入器皿中，撒上花生、香菜即可。

Use This Pot

Soup Pot Or Frying Pan

Ingredients

200g shrimp, 8 mussels, 1 apple, 8 cherry tomatoes, 100g green papaya, 1 red chili pepper, 2T roasted or cooked peanuts, cilantro as needed

Dressings

1 1/2T tamarind sauce, 3T lemon juice, 3T fish sauce, 1 1/2T coconut confection (palm sugar). 2T minced garlic

Marinade

1T minced lemon grass, 1T minced garlic, 1/3T lemon juice, 1t fish sauce, 1t fine granulated sugar, 1T minced basil

Methods

1. Remove shell from shrimp and retain the tail, slice open from the back and devein. Blanch in boiling water until done, remove and drain. Blanch mussels until done in boiling water, then remove and drain.
2. Marinate shrimp and mussels in marinade for 15 minutes until the flavor is absorbed.
3. Peel green papaya, discard seeds and shred finely. Discard the core from apple and cut into slices, then soak in salt water for 5 minutes, remove and drain. Halve cherry tomatoes. Shred red chili pepper. Chop peanuts finely.
4. Combine the ingredients in dressings, green papaya, apples, and cherry tomatoes with dressing. Mix well and let sit for 8 minutes, then add shrimp, mussels and red chili pepper and mix. Transfer to a serving plate, then sprinkle with peanuts and cilantro. Serve.

CC 烹調祕訣 ▌Cooking tips

1. 魚露因品牌不同鹹度略有差異，食譜中的量僅供參考，可以稍微測試。
2. 香茅外層的纖維較粗，這道料理是使用了較軟嫩的香茅內層。
1. There are many fish sauce brands, all of which have different levels of saltiness. The amount of fish sauce in the recipe is just for reference, taste before seasoning.
2. The outer layer of lemon grass is coarser than the inner layer. This recipe uses the tender inner layer.

公主
青豆濃湯

Princess Pea Soup Farm

有別於一般清湯和濃湯，我在這道
以青豆泥為基底的濃湯中加入了白
粥，營養美味，當作主食或湯品都
適合。

To make this different from other, more
ordinary soups, I use white porridge in this
pea paste based thick soup. It can be
served as a main dish or a soup.

使用這個鍋一鍋到底
湯鍋

材料

青豆仁 300 克、培根 6 片、法國麵包丁 1 杯、鮮奶 1 杯、白粥 3 碗、鮮奶油 1/2 杯、基本高湯 1,000c.c.（做法參照 p.6）

調味料

細砂糖 1/3 大匙、粗粒黑胡椒粉適量、鹽適量

做法

1. 先將青豆仁煮熟後，連同鮮奶一起倒入調理機中打成泥狀，然後將白粥和鮮奶油也一起倒入調理機中打成泥狀。
2. 將青豆仁泥和白粥泥倒入鍋中，以小火煮滾，加入調味料拌勻。
3. 培根切絲，倒入鍋中煎炒至變得酥脆，取出瀝掉油分。
4. 將麵包丁放入平底鍋，以小火煎炒至酥脆，或放入烤箱以上下火 120℃烤至酥脆。
5. 將做法 2. 盛入器皿中，撒入適量的培根和麵包丁即可享用。

Use This Pot

Soup Pot

Ingredients

300g peas, 6 slices bacon, 1C French bread croutons, 1C milk, 3 bowls white porridge, 1/2C whipping cream, 1000c.c. basic soup broth (see methods on p.6 for reference)

Seasonings

1/3T fine granulated sugar, coarsely ground black pepper and salt as needed

Methods

1. Cook peas until done first, then blend in a blender with milk added to make paste. Remove, then blend white porridge and whipping cream in blender until a fine paste is created.
2. Combine pea paste and and porridge paste in pan, bring to a boil over low heat, then add seasonings to taste.
3. Shred bacon and fry in pan until crispy, remove and drain.
4. Place diced French bread in oven, bake with upper and lower element at 120℃ until crispy, or fry in frying pan over low heat until crispy.
5. Remove method 2. to serving bowl, sprinkle with bacon and croutons. Yum!

CC 烹調祕訣 | **Cooking tips**

1. 超市有賣冷凍的熟青豆仁，可節省製作的時間。
2. 在濃湯中加入些許細砂糖，可增加甘甜味。
1. Frozen peas can be purchased from the supermarket to save preparation time.
2. Add a little granulated sugar to the thick soup to enhance the sweetness of the soup.

法式鮮蝦蟹肉南瓜湯

French Style Pumpkin with Seafood Soup

這是一道具有豐富層次感，值得推薦給大家的宴客首選湯品。自然清甜的南瓜、肥美多汁的海鮮，精選陸地與海洋的新鮮食材，簡單就能享受到天然的美味。

This dish with its rich layers of taste is worth recommending to everyone as a main soup course. It is very simple, with an enjoyable natural flavor based on the light, sweet, and thick pumpkin, and juicy seafood, along with fresh ingredients from land and sea.

使用這個鍋一鍋到底
湯鍋

材料

螃蟹 2 隻、蝦肉 200 克、蝦仁（帶尾巴）100 克、南瓜 600 克、馬鈴薯 200 克、胡蘿蔔 150 克、洋蔥 1 個、白飯 1 碗、白酒 100c.c.、綜合高湯 1,500c.c.（做法參照 p.9）、牛奶 100c.c.、鮮奶油 100c.c.、奶油 1 大匙

調味料

薑母粉 1 小匙、香蒜粉 1 小匙、鹽適量

做法

1. 螃蟹煮熟後取出蟹肉，蟹殼放於一旁；蝦肉切丁；南瓜去皮去籽後切小丁；馬鈴薯和胡蘿蔔去皮後切小丁；洋蔥切末。

2. 鍋燒熱，倒入奶油，待融化後先加入洋蔥炒至透明，再加入南瓜、馬鈴薯和胡蘿蔔，以小火炒約 5 分鐘，接著加入白飯煮，倒入高湯，煮至白飯和蔬菜軟爛，熄火稍微放涼，然後倒入調理機中攪打成泥狀。

3. 鍋洗淨後燒熱，放入蝦肉丁、蝦仁略炒幾下，淋入白酒。

4. 將做法 2. 倒入湯鍋內，加入牛奶、鮮奶油、蟹肉和做法 3.，加入調味料拌勻，煮至沸騰即可熄火。

這樣做更省時
Time-saving method

製作高湯時，如果使用壓力鍋的話，只要熬煮 1 小時。

It takes only one hour to prepare this soup broth if a Duromatic is used.

Use This Pot
Soup Pot

Ingredients

2 crabs, 200g shrimp meat, 100g shrimp with tails, 600g pumpkin, 200g potato, 150g carrot, 1 onion, 1 bowl white cooked rice, 100c.c. white wine, 1500c.c. soup broth combo (see methods on p.9 for reference), 100c.c. milk, 100c.c. whipping cream, 1T butter

Seasonings

1 ginger powder, 1 garlic power, salt as needed

Methods

1. Cook crabs until done and remove the crab meat. Set the crab shells aside for later use. Dice shrimp meat. Remove skin and seeds from pumpkin, then dice the pumpkin. Peel potato and carrot, then dice. Mince onion.

2. Melt butter in pan, stir-fry onion until transparent, then and pumpkin, potato, and carrot. Saute over low heat for approximately 5 minutes. Add white rice to mix. Pour in soup broth and cook until rice and vegetables are softened. Remove from heat to cool, then blend in blender until thick and mashed.

3. Rinse and heat up the pan, stir-fry diced shrimp and shelled shrimp for a second, then drizzle with white wine.

4. Pour method 2. into the soup pot along with milk, whipping cream, crab meat and method 3.. Season with seasonings to taste. Bring to a boil and remove from heat. Serve.

CC 烹調祕訣 | Cooking tips

1. 這道高湯可用在煮海鮮湯、海鮮麵或海鮮粥、海鮮燉鍋的湯底。
2. 濃湯需以小火煮，以免燒焦。夏天可煮稀一點，冬天則可稍微濃稠。
1. This soup broth can be used as foundation in cooking seafood soup.
2. This thick soup must be prepared over low heat to prevent it from burning. The soup can be a little thinner in the summer and slightly thicker in the winter.

美式
豬肉蔬菜湯
Pork & Veggie Soup

以大量的蔬菜熬煮，湯汁既甘甜且營養，軟嫩的豬肉口感極佳。只要再準備2片外酥脆內柔軟的法國麵包，一頓豐盛的輕食餐點立刻上桌。

If you use lots of different vegetables, the broth turns out to be sweet and nutritious. The pork will be tender with an amazing texture. You only have to prepare two slices of French bread, crispy on the outside and soft on the inside. A rich, delicious light meal will be on the table in a second.

使用這個鍋一鍋到底
湯鍋

材料
梅花肉 600 克、南瓜 400 克、洋蔥 1 個、青豆仁 120 克、肉豆罐頭 1 罐、月桂葉 1 片、白酒 100c.c.、基本高湯 1,500c.c.（做法參照 p.6）

醃料
薑末 1 大匙、香蒜粉 1/2 小匙、粗粒黑胡椒粉、鹽（採用黃金比例）

調味料
黑胡椒粉適量、鹽適量

做法
1. 梅花肉切成 3×3 公分的塊狀；南瓜、洋蔥都切塊狀。
2. 梅花肉撒上醃料，放置 30 分鐘，然後放入鍋中煎至呈金黃色，淋入白酒煮至酒精揮發。
3. 將梅花肉放入湯鍋中，先倒入高湯，續入洋蔥、月桂葉煮約 30 分鐘，再放入南瓜、肉豆和青豆仁，煮至南瓜變軟，加入調味料拌勻即可享用。

Use This Pot
Soup Pot

Ingredients
600g shoulder pork, 400g pumpkin, 1 onion, 120g peas, 1 can pork and beans, 1 bay leaf, 100c.c. white wine, 1500c.c. basic soup broth (see methods on p.6 for reference)

Marinade
1T minced ginger, 1/2t garlic powder, coarsely ground black pepper and salt as needed (use golden proportion)

Seasonings
coarsely ground black pepper and salt as needed

Methods
1. Cut shoulder pork into 3cmx3cm big pieces. Cut pumpkin and onion into pieces.
2. Marinate shoulder pork in marinade for 30 minutes until the flavor is absorbed. Fry in pan until golden and drizzle with white wine, cook until the alcohol evaporates.
3. Remove pork to pot, pour in soup stock first, then onions and bay leaf. Cook for about 30 minutes, then add pumpkin, pork and beans as well as peas. Continue cooking until pumpkin turns soft. Season with seasonings to taste, stir to mix. Ready to serve.

CC 烹調祕訣 ▌Cooking tips

薑和南瓜味道很合，此外，也可以用薑母粉，一樣美味。
Ginger goes well with pumpkin. Ginger powder may be used instead of whole ginger if desired.It's also delicious.

西西里島海鮮湯

Sicily Seafood Soup

看著鍋中豪華無比的海鮮食材，忍不住要流口水囉！只要事先熬好高湯，這道看似困難的湯料理其實做法很簡單。

Just looking at the rich seafood ingredients in the pot already makes my mouth water. This soup recipe seems difficult, yet if the soup broth is prepared beforehand, it can be very simple.

使用這個鍋一鍋到底
湯鍋

材料

去骨魚肉 200 克、蛤蜊 200 克、貽貝 8 個、大蝦 4 尾、小蝦 16 尾、透抽 200 克、小花枝 8 隻、蕃茄粒罐頭 1 罐、蒜末 3 大匙、洋蔥末 2 大匙、綜合高湯 1,500c.c.（做法參照 p.9）、白酒 150c.c.、月桂葉 1 片、羅勒適量、法國麵包適量、橄欖油 2 大匙

調味料

粗粒黑胡椒粉適量、鹽適量

做法

1. 魚肉切大塊；蛤蜊放入鹽水中浸泡，使其吐沙；透抽切圈狀；蕃茄粒搗碎。
2. 鍋中倒入 1 大匙橄欖油，待油熱後放入 2 大匙蒜末爆香，續入洋蔥末炒至呈透明，加入蕃茄粒，倒入高湯，放入月桂葉，改小火維持沸騰狀態煮約 10 分鐘。
3. 另取一鍋，倒入 1 大匙橄欖油，待油熱後放入 1 大匙蒜末爆香，續入蛤蜊煮至殼微開，再加入貽貝、大蝦、魚肉、小蝦和透抽、小花枝，淋入白酒煮至酒精揮發，再倒入做法 2.，加入調味料拌勻，撒上羅勒。
4. 盛入器皿中，搭配酥脆的法國麵包最好吃！

Use This Pot
Soup Pot

Ingredients

200g boneless fish, 200g clams, 8 mussels, 4 large shrimp, 16 small shrimp, 200g squid, 8 small squids, 1 can whole tomato, 3T minced garlic, 2T minced onion, 1500c.c. soup broth combo (see methods on p.9 for reference), 150c.c. white wine, 1 bay leaf, sweet basil as needed, French bread as needed, 2T olive oil

Seasonings

coarsely ground black pepper and salt as needed

Methods

1. Cut fish meat into large pieces. Soak clams in salt water to let them spit out any impurities.Squid cut into round sections. Crush tomatoes.
2. Heat 1T olive oil in pot until smoking, add 2T minced garlic until fragrant. Fry minced onion until flavor is released. Add whole tomatoes, pour soup broth in, add bay leaf, and continuing boiling for about 10 minutes.
3. Take another pot, pour 1T olive oil and heat until smoking, add 1T minced garlic until fragrant, then clams and cook until shells open. Put mussels, large shrimp, fish meat, small shrimp, and squid, small squid in, drizzle in white wine and cook until the alcohol evaporates, then pour in the ingredients from method 2. to mix well. Sprinkle with basil.
4. Place it in a serving plate and pair with crispy French bread for a more delicious meal. Serve!

CC 烹調祕訣 | Cooking tips

1. 將麵包移入烤箱，以上下火 140℃烤至酥脆或用平底鍋以小火煎炒至酥脆。
2. 貽貝又叫淡菜、孔雀貝，香港人則稱為青口。
1. Put the bread in an oven and heat with upper or lower element at 140℃, or heat it in frying pan over low heat until crispy.
2. Mussels have many local names in the languages of China. For example, in Hong Kong they are called ching kou.

東京風味
豆漿什錦鍋

Tokyo Flavored
Soy Bean Milk Hot Pot

這鍋4～6人份的營養什錦養生湯，是以無糖豆漿為鍋底，再加入許多食蔬。在吃美食的同時，又能攝取豐富的營養、大量的纖維，健康又美味，很受女性歡迎。

This nutritious healthy soup recipe serves 4 to 6 people. The foundation of the soup is unsweetened soy bean with lots of vegetables. This delicious dish is packed with nutrition and fiber. Both healthy and tasty, it is very popular with females

使用這個鍋一鍋到底
湯鍋

材料

去骨雞腿 2 隻、火鍋肉片（豬肉或牛肉）200 克、蝦子 150 克、蛤蜊 200 克、豆腐 1 塊、新鮮香菇 8 朵、金針菇 1 包、芹菜 80 克、無糖豆漿 1,000c.c.、日式高湯（做法參照 p.8）600c.c.、芝麻葉 80 克、烏龍麵適量、蔥 3 支

醃料

味噌 1 大匙、香油 1/2 大匙、醬油 1 1/2 大匙、味醂 2/3 大匙、蔥末 1 大匙

沾醬

柴魚醬油 5 大匙、味醂 1 大匙、檸檬汁 2 1/2 大匙、香橙皮末 1/2 大匙、蔥末 3 大匙

調味料

鹽適量

做法

1. 先將醃料拌勻。
2. 雞腿肉切塊，放入醃料中拌勻，放置 1 小時。豆腐切小塊；芹菜切段；蔥切絲。
3. 將豆漿、高湯倒入鍋中煮沸，加入調味料，放入雞腿肉、肉片、蝦子、蛤蜊、豆腐、香菇、金針菇、芹菜煮熟，最後放芝麻葉、烏龍麵、蔥絲即可上桌。
4. 製作沾醬：將沾醬的材料拌勻即可。
5. 食用時，可搭配沾醬一起享用。

Use This Pot
Soup Pot

Ingredients

2 boneless chicken legs, 200g hot pot pork slices (pork or beef), 150g shrimp, 200g clams, 1 piece of square tofu, 8 fresh shiitake mushrooms, 1 pack Enoki mushrooms, 80g celery, 1000c.c. unsweetened soy bean milk, 600g Japanese soup broth (see methods on p.8 for reference), 80g sesame seed leaves, udon noodles as desired, 3 scallions

Marinade

1T miso paste, 1/2T sesame oil, 1 1/2T soy sauce, 2/3T mirin, 1T minced scallions

Dipping Sauce

5T bonito flavored soy sauce, 1T mirin, 2 1/2T lemon juice, 1/2T minced orange peel, 3T shredded scallions

Seasonings

salt as needed

Methods

1. Combine all ingredients in Marinade well together.
2. Cut chicken into pieces and marinate in marinade for 1 hour until the flavor is well absorbed. Cut tofu into small pieces. Cut celery into sections. Shred scallions.
3. Place tofu and soup broth in cooking pan with seasonings added. Then put in chicken, pork, shrimp, clam, tofu, Enoki mushrooms, celery, and sesame seed leaves along with the udon noodle and shredded scallions. Remove and place on serving table.
4. To prepare dipping sauce: Combine all ingredients in dipping sauce well together.
5. Serve with the dipping sauce on the side as a dip.

CC 烹調祕訣 | Cooking tips

刨香橙皮時，要避開皮白色的部分，否則會有苦味。

Avoid the white part of the orange when grating the orange skin, or it will taste bitter.

料理之外，你和我都還有另一個胃，
即便不是時時刻刻需要，
卻也日日不能少。
華麗的什錦水果煎餅是假日的夢幻早午餐，
優雅的香蕉巧克力蛋糕讓你歡欣品嚐下午茶，
樸實的加州風味漢堡原味食材令人滿足，

準備好了嗎？
來一場點心大冒險吧！

Taiwanese often say when dessert arrives: "Oh my! I need another stomach!" Even if we do not need it all the time, we cannot live without it. The rich Champs-Elysee Fruit-mix Pancake makes a dream brunch for the holidays, while the exquisite English Banana Chocolate Cake will bring joy to your afternoon tea. The ingredients in this simple California Cheeseburger fill you with satisfaction. Are you ready for a dessert adventure?

PART 4
點心和麵包類
DESSERTS AND BREADS SECTION

英式
香蕉巧克力蛋糕
English Banana Chocolate Cake

家中沒有烤箱卻很想做蛋糕的人，可以試試這道用平底鍋做的香蕉巧克力蛋糕。口感鬆軟、香蕉風味濃厚，加入苦甜巧克力更顯獨特。

I am a person who likes to make cake but there is no oven in my house. Hence, I try making this delicious banana chocolate cake with a frying pan. The taste and texture are soft and the banana flavor strong. The bittersweet chocolate makes it unique.

使用這個鍋一鍋到底
平底鍋

材料
58% 苦甜巧克力 200 克、無糖鮮奶油 100 克、香蕉 4 條、全蛋 4 個、奶油 80 克、君度橙酒 30c.c.、低筋麵粉 200 克、泡打粉（可不放）1 小匙、糖粉適量

做法
1. 將巧克力放入鍋內，倒入鮮奶油，鍋底隔一盆水，隔水加熱，以小火並用木杓拌勻至巧克力融化產生透亮光澤。
2. 香蕉去皮後搗成泥狀；全蛋攪拌均勻；低筋麵粉和泡打粉過篩。
3. 奶油放在室溫下使其軟化，然後倒入盆中，倒入巧克力鮮奶油、香蕉泥、君度橙酒和蛋液輕輕拌勻。
4. 加入低筋麵粉、泡打粉輕輕拌勻成麵糊，不可拌太久。
5. 將麵糊倒入鍋內，蓋上鍋蓋，以小火烤約 30 分鐘，取出蛋糕，待蛋糕涼後撒上糖粉即可。

這樣做更省時
Time-saving method

做法 1. 中融化巧克力鮮奶油時，若使用瑞康屋的平底鍋，不需隔水，直接用木杓開小小火拌煮至巧克力融化產生透亮光澤即可。

In method 1. melting the chocolate and whipping cream, if a Rikon suate pan is used, there is no need to put a pan inside a pan and fill with water. Just heat on low and stir with the wooden spoon until the chocolate melts and the color of both ingredients is smoothly combined.

Use This Pot
Frying Pan

Ingredients
200g 58% semisweet chocolate, 100g whipping cream, 4 bananas, 4 eggs, 80g butter, 30c.c. cointreau, 200g low gluten flour, 1t baking powder (optional), sugar powder as needed

Methods
1. Put chocolate and whipping cream in a pan, then remove the pan to a bigger pan full of water, then cook over low heat and stir constantly with a wooden spoon until dissolved and the color is bright and shiny.
2. Remove skin from banana and mash. Beat eggs until evenly mixed. Sift cake flour and baking powder.
3. Place butter at room temperature until soften, then remove to a mixing bowl along with chocolate whipping cream, mashed banana, cointreau, and egg. Stir gently until evenly-mixed.
4. Fold in cake flour and baking soda, keep on stirring gently until the batter is smooth and even. (Do not stir for too long.)
5. Pour the batter in frying pan and cover. Bake at low heat for about 30 minutes until done, then remove to cool and sprinkle with sugar powder on top. Ready to serve.

CC 烹調祕訣 | Cooking tips
融化巧克力時的溫度不可超過 38℃，否則巧克力易變質。此外，使用木杓或木匙拌勻為佳，使用不銹鋼材質杓或匙的話，溫度傳熱過快。

The temperature cannot exceed 38℃ when melting the chocolate, or it will easily burn. Stir with a wooden spoon rather than steel, which conducts heat to fast.

香榭麗舍
什錦水果煎餅

Champs-Elysées
Fruit-Mix Pancake

這道料理是我的口袋私房甜點之一，很適合在小朋友的聚會、下午茶或飯後甜點食用。煎餅用途很廣，可包鮪魚、雞肉等鹹餡料或者巧克力冰淇淋甜品。

This is one of my confidential dessert recipes. It is perfect for a children's party, afternoon tea, or an after meal dessert. The pancakes have a great many uses. They may be stuffed with salty ingredients, such as tuna, chicken, or with sweet desserts, such as chocolate or ice cream.

使用這個鍋一鍋到底
平底鍋

材料
蘋果 1 個、柳丁 1 個、哈密瓜 1/2 個、香蕉 1 條、君度橙酒 2 大匙、細砂糖 1 大匙、新鮮薄荷適量、奶油 3 大匙

麵糊
低筋麵粉 1 杯、牛奶 1 杯、全蛋 2 個、玉米粉 3 大匙、柳丁皮末 1/2 大匙、檸檬皮末 1/2 大匙、融化的奶油 1 大匙、白砂糖 2 1/2 大匙、鹽 1 小匙

做法
1. 將麵糊中的低筋麵粉過篩後，與麵糊中的其他材料一起倒入盆中，以打蛋器拌勻呈濃稠狀的麵糊。
2. 蘋果去皮去核後切塊，放入鍋中，倒入 1 大匙細砂糖、1 大匙奶油煮約 3 分鐘，淋上 1 大匙君度橙酒拌勻。
3. 柳丁取出果肉；哈密瓜和香蕉都切塊。
4. 鍋中倒入 1 大匙奶油，待奶油融化後放入柳橙、香蕉拌勻，再淋入 1/2 大匙君度橙酒拌勻。哈密瓜放入盆中，淋入 1/2 大匙君度橙酒拌勻。
5. 鍋面刷入少許奶油，待油熱後，倒入適量的麵糊煎成圓形餅狀，取出排入盤中。
6. 將水果包在餅皮中，捲成長筒狀，最後以水果、薄荷裝飾即可。

Use This Pot
Frying Pan

Ingredients
1 apple, 1 orange, 1/2 cantaloupe, 1 banana, 2T cointreau, 1T fine granulated sugar, fresh mint leaves as needed, 3T butter

Batter
1C cake flour, 1C milk, 2 eggs, 3T cornstarch, 1/2T minced orange peel, 1/2T minced lemon zest, 1T melted butter, 2 1/2T white granulated sugar, 1t salt

Methods
1. Sift the cake flour first, then mix well with the rest of the ingredients, beat with an egg beater until batter is formed.
2. Peel apple, discard core and cut into pieces, then place in pan with 1T of fine granulated sugar and 1T of butter added. Cook for approximately 3 minutes, drizzle with 1T of cointreau to mix.
3. Remove orange flesh. Cut banana and cantaloupe into pieces.
4. Heat 1T of butter until melted, add orange and banana to mix. Drizzle 1/2T of cointreau and mix well. Place cantaloupe in bowl and drizzle with 1/2T of cointreau to mix.
5. Brush butter over to grease pan, scoop suitable amount of batter in pan and fry into a round pancake. Continue the process to finish the batter, then remove to serving plate.
6. Wrap the fruit inside the pancake and roll up into a cylinder. Garnish with fruit and mint leaves. Serve.

CC 烹調祕訣 Cooking tips

也可以使用奇異果、桃子和梨子取代蘋果、哈密瓜等。
Kiwi, peach and pear may be used in place of apple, cantaloupe.

義大利
南瓜乳酪蛋糕
Italian Pumpkin Cheese Cake

這道做法簡單的蛋糕可以說是我的招牌甜點，由於顏色討喜，過年時拿來招待客人最喜氣，我則稱它為「黃金乳酪蛋糕」。

This simple and easy to make cake is my specialty dessert. Because of its rich color, I always make this during the New Year holiday to entertain guests. I call it golden cheese cake.

使用這個鍋一鍋到底
平底鍋

材料

南瓜 400 克、奶油乳酪 200 克、奶油 1 小條、細砂糖 120 克、全蛋 4 個、低筋麵粉 300 克、泡打粉 1/2 大匙、香草粉 1/2 小匙、鹽 1/2 小匙、油少許

做法

1. 南瓜去皮去籽，放入鍋中蒸熟，取出搗成泥；乳酪奶油、奶油於常溫下放至軟化。
2. 奶油放入盆內，倒入細砂糖拌勻，再加入蛋拌勻，然後加入南瓜、乳酪奶油，以打蛋器拌勻。
3. 低筋麵粉、泡打粉和香草粉分別過篩後倒入做法 2. 中，輕輕拌勻成麵糊。
4. 鍋中抹上少許油，倒入麵糊，蓋上鍋蓋，以小火加熱 30 分鐘即可。

Use This Pot
Frying Pan

Ingredients

400g pumpkin, 200g cream cheese, 1 small bar butter, 120g fine granulated sugar, 4 eggs, 300g cake flour, 1/2T baking powder, 1/2t vanilla powder, 1/2t salt, cooking oil as needed

Methods

1. Remove skin and seed from pumpkin, heat it in pan until done, then remove and mash finely. Let cream cheese and butter sit at room temperature until softened.
2. Put butter in a mixing bowl, add granulated sugar to mix, then egg as well as pumpkin and cream cheese, beat with an egg beater until well-mixed.
3. Sift cake flour, baking powder, and vanilla powder separately first, then add to method 2., stir gently and lightly to form batter.
4. Grease pan with oil as needed, pour batter in, cover and simmer over low heat for 30 minutes until done. Serve.

CC 烹調祕訣 ▍Cooking tips

1. 建議使用味道佳的東昇南瓜。東昇南瓜又名栗子南瓜，肉質結實，嘗起來像栗子，口感較 Q。
2. 做法 2 中攪拌時，要每一樣食材拌勻後才能加入下一樣食材，以免成品口感有顆粒。
1. Dongsheng pumpkin, also known as chestnut pumpkin, is suggested for use in this recipe because of its sweet chestnut taste and firm texture.
2. In method 2, when ingredients are added to the mixing bowl, make sure the batter is well-mixed before adding the next ingredient, or there will be lumps in the cake.

加州風味乳酪漢堡

California Cheeseburger

這款深受大人小孩歡迎的4人份乳酪漢堡，多樣的蔬菜和絞肉，營養滿點。假日的早上，不妨來一個大漢堡當早午餐吧！

This wonderful cheeseburger recipe is the one that both adults and children can fully enjoy! It has all sorts of vegetables and ground meat with lots of nutrition. During holidays, why not make yourself a burger for breakfast or lunch?

使用這個鍋一鍋到底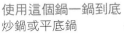
炒鍋或平底鍋

材料

絞肉 600 克、洋蔥 1 個、麵包粉 5 大匙、鮮奶 100c.c.、蛋液 1 個份量、雞蛋 4 個、麵粉適量、乳酪適量、油適量

調味料

粗粒黑胡椒粉 1/2 大匙、鹽 1 大匙、荳蔻粉 1/3 小匙

醬料

牛蕃茄 2 個、檸檬汁 1 大匙、鹽少許、粗粒黑胡椒粉少許、橄欖油 1 大匙、香菜末 1 大匙、蒜末 1/2 大匙、洋蔥末 1 大匙

做法

1. 洋蔥切末；麵包粉浸泡鮮奶約 5 分鐘；牛蕃茄切小丁，和醬料中的所有材料拌勻。
2. 鍋中倒入油，待油熱後放入洋蔥炒至呈透明，放涼。
3. 將絞肉放入調理盆中，倒入做法 2. 的洋蔥、浸泡過鮮奶的麵包粉、調味料、蛋液，拌勻至有點黏稠狀，然後取一份絞肉壓成圓扁狀，沾上麵粉。
4. 鍋中倒入油，待油熱後放入絞肉煎熟，取出放在器皿中，先鋪上醬料，再放上乳酪，以小火加熱至乳酪融化。
5. 煎好一個荷包蛋，放在器皿中，再放上做法 4. 即可享用。

Use This Pot
Wok Or Frying Pan

Ingredients
600g ground meat, 1 onion, 5T bread crumb, 100c. c. milk, 1 egg liquid proportion, 4 chicken eggs, flour as needed, cheese as needed, cooking oil as needed

Seasonings
1/2T coarsely ground black pepper, 1T salt, 1/3t nutmeg powder

Sauce
2 beefsteak tomatoes, 1T lemon juice, salt as needed, coarsely ground black pepper as needed, 1T olive oil, 1T minced cilantro, 1/2T minced garlic, 1T minced onion

Methods
1. Mince onion. Soak bread crumbs in milk for about 5 minutes. Chop the beefsteak tomato into small dices. Combine the ingredients from the sauce well.
2. Heat oil in pan until smoking, fry onion until flavor is strong, place in cool.
3. Combine ground pork in mixing bowl with onions from method 2., bread crumbs soaked in milk, seasonings and egg, stir until thick and sticky. Scoop up some pork mixture in your hand, pat and press to form a flat patty, then coat evenly with flour.
4. Heat oil in pan until smoking, fry patty until done, then place in pan. Spread the sauce over, add cheese, then heat over low until cheese melts.
5. Fry a sunny sideup egg and place it on the serving plate, then top with patty from method (4). Serve.

CC 烹調祕訣 | Cooking tips

麵包粉浸泡牛奶時，牛奶不宜過多，只要讓麵包粉浸泡到牛奶即可。

Do not pour in too much milk when soaking the bread crumbs, just enough to soak the bread crumbs.

希臘鮮蝦麵包

Greek Shrimp with Croutons

以西式料理宴客時前菜絕對不可少，它不僅可以讓賓客在食用主菜前先墊墊肚子，更肩負開胃的重責大任，千萬別輕忽了喔！

The appetizer is a must in a western banquet. It not only fills the guest's stomachs before serving the main course, it also stimulates their appetites. It is so important that you cannot simply ignore it.

使用這個鍋一鍋到底
湯鍋或平底鍋

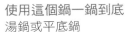

材料
法國麵包丁 1 杯、羅蔓生菜 1 顆、蝦子 300 克、洋蔥末 2 大匙、蒜末 1 大匙、檸檬汁 2 大匙、牛蕃茄 2 個、炸好的酥脆蒜片 3 大匙、香菜末 2 大匙、橄欖油 2 大匙、香蒜粉 1 大匙

調味料
粗粒黑胡椒粉適量、鹽適量

做法
1. 生菜淨後放入冰水中冰鎮 20 分鐘，取出瀝乾水分；牛蕃茄切丁。
2. 蝦子汆燙熟，留下幾尾去殼留尾巴，其餘的都切成蝦肉丁。
3. 將蝦肉丁放入盆中，倒入洋蔥末、蒜末、牛蕃茄、酥脆蒜片、香菜末、橄欖油和調味料拌勻，放置 10 分鐘。
4. 麵包丁撒上香蒜粉，移入烤箱，以上下火 140℃烤至酥脆或平底鍋以小火煎炒至酥脆。將麵包丁搭配做法 **3.** 的料一起食用。
5. 將做法 **3.** 鋪在生菜上，撒入酥脆的蒜片和做法 **4.** 的麵包丁，最後擺上整尾的蝦即可享用。

Use This Pot
Soup Pot Or Frying Pan

Ingredients
1C croutons, 1 head fresh romance lettuce, 300g shrimp, 2T minced onion, 1T minced garlic, 2T lemon juice, 2 beefsteak tomatoes, 3T deep-fried garlic slices, 2T minced cilantro, 2T olive oil, 1T garlic powder

Seasonings
coarsely ground black pepper and salt as needed

Methods
1. Rinse lettuce leaves well and soak in ice water for 20 minutes, then remove and drain. Cut beefsteak tomatoes into pieces.
2. Blanch shrimp until done, leave a few shrimp with tails on, and dice the rest of the shrimp finely.
3. Place the diced shrimp in mixing bowl with minced onion, minced garlic, beefsteak tomato, deep-fried garlic slices, cilantro, olive oil, and ingredients from seasonings added. Stir to mix, then let sit for 10 minutes until flavor is absorbed.
4. Sprinkle garlic powder over croutons and bake in oven with upper or lower element at 140℃ until crispy, or fry in frying pan over low heat until crispy. Serve with shrimp from method **3.** on the side. Serve.
5. Line method **3.** evenly on the lettuce, sprinkle with crispy garlic slices and croutons from method **4.**, then place a whole shrimp on top. Serve.

CC 烹調祕訣 | Cooking tips

牛蕃茄切成丁後，必須仔細將水分瀝乾再使用。
After dicing the beef steak tomato, the liquid has to be carefully drained.

韓國泡菜海鮮煎餅

Korean Kimchi Seafood Pan Cake

可當主食、點心的海鮮煎餅，是我
吃韓國料理時必點的一道菜，可隨
個人喜好加入海鮮食材。

This seafood pancake dish can be served as a
main course. It is one of my favorite dishes and I
always have it whenever I eat Korean food. Any
seafood can be added to this dish as desired.

使用這個鍋一鍋到底
平底鍋

材料

低筋麵粉 3/4 杯、糯米粉 1/4 杯、泡菜 150 克、透抽 150 克、蝦仁 150 克、蔥 4 支、雞蛋 2 個、冷的綜合高湯（做法參照 p.9）、1 1/4 杯、油 1/2 大匙、香油 1 大匙

調味料

鹽適量、米酒 1/2 大匙

沾醬

韓國辣椒醬 2 大匙、淡色醬油 2 大匙、味醂 1 大匙、檸檬汁 1 大匙、香油 1 大匙、熟白芝麻 1/2 大匙、薑泥 1/3 大匙、蔥花 1 大匙、基本高湯 2 大匙（做法參照 p.6）

做法

1. 泡菜切絲；蔥切段；蝦仁由背部劃開，攤平取出腸泥，洗淨瀝乾水分；透抽切條，洗淨瀝乾水分。
2. 低筋麵粉過篩後倒入盆中，放入糯米粉，先慢慢倒入冷高湯拌勻，再倒入蛋液拌勻，然後加入做法 1. 和調味料拌勻成煎餅糊，放置 30 分鐘。
3. 鍋中倒入油和香油，待油熱後倒入適量的煎餅糊，煎至兩面都呈金黃色。
4. 將沾醬的所有材料拌勻，再搭配煎餅一起食用。

Use This Pot
Frying Pan

Ingredients
3/4 cake flour, 1/4C sticky rice flour, 150g kimchi, 150g squid, 150g shrimp, 4 scallions, 2 eggs, 1 1/4C cold soup broth combo (see methods on p.9 for reference), 1/2T cooking oil, 1T sesame oil

Seasonings
Salt as needed, 1/2T rice wine

Dipping Sauce
2T Korean chili sauce, 2T light soy sauce, 1T mirin, 1T lemon juice, 1T sesame oil, 1/2T roasted white sesame seeds, 1/3T mashed ginger, 1T chopped scallions, 2T basic soup broth (see methods on p.6 for reference)

Methods
1. Shred kimchi. Cut scallions into sections. Slice shrimp open from the back, spread flat and remove any impurities, then rinse well and drain. Cut squid into strips, rinse well and drain.
2. Sift flour and remove a mixing bowl. Add sticky rice flour, then pour in cold soup broth slowly and mix well. Add eggs and stir well. Add method 1. and seasonings to make pancake batter, then let rest for 30 minutes.
3. Heat cooking oil and sesame oil in pan until smoking, then pour in suitable amount of batter, fry until golden on both sides.
4. Combine all the ingredients from Dipping Sauce well together and serve on the side of the pancake.

CC 烹調祕訣 | **Cooking tips**

1. 低筋麵粉的顆粒較粗，一定要過篩後再使用，成品口感較佳。
2. 搭配煎餅的沾醬還可以用來拌乾麵或做涼拌料理。
1. Cake flour is coarse in texture, sift first to guarantee a better texture.
2. The dipping sauce can also be used in mixing with dried noodles or cold dishes.

一個鍋做異國料理
全世界美食一鍋煮透透（中英對照）

作者	洪白陽（CC 老師）
攝影	廖家威
翻譯	施如瑛
美術設計	黃祺芸
編輯	彭文怡
行銷	呂瑞芸
企畫統籌	李橘
總編輯	莫少閒
出版者	朱雀文化事業有限公司
地址	台北市基隆路二段 13-1 號 3 樓
電話	（02）2345-3868
傳真	（02）2345-3828
劃撥帳號	19234566　朱雀文化事業有限公司
e-mail	redbook@ms26.hinet.net
網址	http://redbook.com.tw
總經銷	成陽出版股份有限公司
ISBN	978-986-6029-49-3
初版一刷	2013.11

定價	350 元
出版登記	北市業字第 1403 號

About 買書：

●朱雀文化圖書在北中南各書店及誠品、金石堂、何嘉仁等連鎖書店均有販售，如欲購買本公司圖書，建議你直接詢問書店店員。如果書店已售完，請撥本公司經銷商北中南區服務專線洽詢。北區（03）358-9000、中區（04）2291-4115 和南區（07）349-7445。

●●至朱雀文化網站購書（http://redbook.com.tw），可享 85 折起優惠。

●●●至郵局劃撥（戶名：朱雀文化事業有限公司，帳號 19234566），

掛號寄書不加郵資，4 本以下無折扣，5～9 本 95 折，10 本以上 9 折優惠。

港澳地區授權出版：Forms Publications (HK) Co. Ltd.
地址：香港北角英皇道 499 號北角工業大廈 18 樓
電話：（852）2138-7998
傳真：（852）2597-4003
電郵：marketing@formspub.com
網站：http://www.formspub.com
　　　http://www.facebook.com/formspub

港澳地區代理發行：香港聯合書刊物流有限公司
地址：香港新界大埔汀麗路 36 號
　　　中華商務印刷大廈 3 字樓
電話：（852）2150-2100
傳真：（852）2407-3062
電郵：info@suplogistics.com.hk
ISBN：978-988-8237-66-1

出版日期：二零一三年十一月第一次印刷

KUHN RIKON
SWITZERLAND
瑞康屋

百貨專櫃據點

台北:
士林旗艦店 1F
新光三越台北南西店 7F
太平洋SOGO百貨復興店 8F
太平洋SOGO百貨忠孝店 8F
統一阪急百貨台北店 6F
新光三越台北信義新天地A8 7F
板橋大遠百Mega City 7F
HOLA特力和樂 士林店 B1
HOLA特力和樂 內湖店 1F
HOLA特力和樂 中和店 1F
HOLA特力和樂 土城店 3F

桃園:
FE21'遠東百貨 桃園店 10F
太平洋SOGO百貨中壢元化館 7F
HOLA特力和樂 南崁店 1F

新竹:
新竹大遠百 5F
太平洋SOGO百貨新竹店 9F
太平洋崇光百貨巨城店 6F

台中:
新光三越台中中港店 8F
HOLA特力和樂 中港店 1F
HOLA特力和樂 北屯店 1F
台中大遠百Top City 9F

台南:
新光三越台南西門店 B1
HOLA特力和樂 仁德店 2F

嘉義:
HOLA特力和樂 嘉義店 1F

高雄:
新光三越高雄左營店 9F
統一阪急百貨高雄店 5F
HOLA特力和樂 左營店 1F

KUHN RIKON
SWITZERLAND

瑞 康 屋

瑞康國際企業股份有限公司 TEL 0800 39 3399 FAX 02 8811 2518 www.rakenhouse.com

One Pan for Exotic Meals

One Pan for Exotic Meals